Canto is an imprint offering a range of
titles, classic and more recent, across a
broad spectrum of subject areas and
interests. History, literature, biography,
archaeology, politics, religion, psychology,
philosophy and science are all represented
in Canto's specially selected list of titles,
which now offers some of the best and
most accessible of Cambridge publishing
to a wider readership.

Erwin Schrödinger was a brilliant and charming Austrian, one of the greatest scientists of the twentieth century, and a man with a passionate interest in people and ideas. He was best known for the discovery of wave mechanics, which won him the Nobel Prize for Physics, but his most influential book, *What Is Life?*, served to attract some of his brightest scientific contemporaries into molecular biology. This abridgement of Walter Moore's award-winning *Schrödinger: Life and Thought* offers a highly readable account, interweaving Schrödinger's scientific work with his intense personal friendships, his interest in mysticism, and the turbulent background of political events in Germany and Austria, where he barely escaped from vindictive Nazis. Hailed as a "breathtaking accomplishment," Walter Moore's colorful and sympathetic account looks set to become a classic of scientific biography.

A LIFE OF ERWIN SCHRÖDINGER

A LIFE OF ERWIN SCHRÖDINGER

WALTER MOORE

CAMBRIDGE
UNIVERSITY PRESS

Published by the Press Syndicate of the University of Cambridge
The Pitt Building, Trumpington Street, Cambridge CB2 1RP
40 West 20th Street, New York, NY 10011-4211, USA
10 Stamford Road, Oakleigh, Melbourne 3166, Australia

First published 1994

Printed in Great Britain at the University Press, Cambridge

A catalogue record for this book is available from the British Library

Library of Congress cataloguing in publication data
Moore, Walter John, 1918–
A life of Erwin Schrödinger / Walter Moore.
p.   cm.
Includes index.
ISBN 0 521 46934 1 paperback
1. Schrödinger, Erwin, 1887–1961.
2. Physicists – Austria – Biography.   I. title.
DC16S265M65   1994
530′.092 – dc20   93–50157   CIP
[B]

ISBN 0 521 46934 1 paperback

Contents

Illustrations

xi

Preface

In this book I have tried to tell something of the life of Erwin Schrödinger in such a way that even those who are not scientists may be able to understand the greatness of his work, the remarkable range of his ideas, and the complex nature of his personality. This Canto edition is a shorter version of *Schrödinger: Life and Thought* (1989), to which those interested are referred for extensive references and documentation.

A biography of Schrödinger was made possible by the generosity with which his eldest daughter, Mrs. Ruth Braunizer, made available many of the archives concerning her father's life.

Of the people who helped in various ways, only a few will be mentioned again in this edition: Professor Ludvik Bass, Dr. Linda Wessels, Professor James McConnell, Professor Bruno Bertotti, Mrs. Hansi Bauer-Bohm, Dr. Wolfgang Kerber, Professor Nicholas Kurti. Special thanks are due to the Hebrew University of Jerusalem for permission to quote from the letters of Albert Einstein, and to Professor Gustav Born for permission to quote from the letters of his father, Max Born.

I have been especially thankful over the years for the friendly advice of Dr. Simon Capelin of the Cambridge University Press, who suggested the preparation of this Canto edition.

The book could not have been written without the help of my wife Patricia who arranged and recorded the many interviews, managed an extensive correspondence, and suggested the most tactful pathways for research.

Introduction

When the First World War began in August 1914, Erwin Schrödinger at the age of twenty-seven was a *Privat Dozent* at Vienna University, at the outset of a promising career in physics. He served throughout the war as an officer in the Austrian Fortress Artillery, first on the Italian front and later assigned to the imperial meteorology department in Vienna. On October 7, 1917, his revered professor of theoretical physics, Fritz Hasenöhrl, was killed by a grenade in an Italian attack on Mount Plaut in the South Tirol.

Early in 1918, Schrödinger received word that he was being seriously considered for the position of *Professor Extraordinarius* (associate professor) in theoretical physics at the University of Czernowitz.

I made up my mind to lecture there on honest theoretical physics, first of all according to the pattern of the splendid lectures of my beloved teacher Fritz Hasenöhrl, fallen in the war, but besides to concern myself with philosophy, deeply immersed as I then was in the writings of Spinoza, Schopenhauer, Mach, Richard Semon, and Richard Avenarius. My good angel intervened, since soon Czernowitz no longer belonged to us. So nothing came of it. I had to stick to theoretical physics, and to my astonishment something occasionally emerged from it.

A serious interest in philosophy was not unusual among German and Austrian physicists, but for Schrödinger at that time it was so important that he was even tempted to sacrifice his scientific research and devote his life to philosophical study. Much later, in 1963, Max Born wrote: "I am convinced that theoretical physics is actually philosophy. It has revolutionized

1

fundamental concepts, e.g., about space and time (relativity), about causality (quantum theory), and about substance and matter (atomistics). It has taught us new methods of thinking (complementarity), which are applicable far beyond physics." By his discovery of wave mechanics in 1926, Erwin Schrödinger contributed in a major way to this revolutionary change in our view of the world. According to Arnold Sommerfeld, second to none in his understanding of modern physics: "It was the most astonishing among all the astonishing discoveries of the twentieth century." This judgment should also include quantum mechanics, which was discovered a year earlier by Werner Heisenberg.

Why were these theories so revolutionary in their philosophical implications? In summary, they had the following results: (1) They devalued the philosophy of materialism that had prevailed in science since the time of Newton. (2) They showed that everything in the world is part of everything else; there are no boundaries and no isolated parts in the world. (3) They destroyed the idea of strict determinism and predictability in nature. (4) They raised deep questions about the role of mind in nature, and thus restored the human observer to a central position in natural philosophy. As a consequence of all these overturnings of conventional ideas, the influence of the new quantum theory spread far beyond physics and left the philosophy of science, and indeed philosophy in general, in a state of flux whose resolution cannot as yet be discerned.

In his book *The Structure of Scientific Revolutions* (1962), Thomas Kuhn argued that there are two kinds of science, normal science and revolutionary science. In any particular field, normal science is conducted in accordance with a set of rules, concepts, and procedures, called a *paradigm*, which is accepted by all scientists working in that field. Normal science is similar to puzzle-solving: interesting, even beautiful, solutions are obtained but the rules are never changed. In this normal scientific activity, however, unexpected discoveries sometimes are made that are inconsistent with the prevailing paradigm. Among the scientists, a tense situation ensues, which increases in intensity until a scientific revolution is achieved.

This is marked by a *paradigm shift*, and a new paradigm emerges under which normal scientific activity can be resumed. This analysis of scientific progress appears to fit the work of Schrödinger. His scientific papers before 1925 were typical products of "normal science," competent but certainly not remarkable extensions of subjects that were being pursued by his teachers and co-workers at the University of Vienna. In 1926, his *annus mirabilis*, he published at the "advanced age" of thirty-eight his four great papers on wave mechanics, of which Born said: "there is nothing more beautiful in theoretical physics." This was revolutionary science by any standard. After his great discovery, Schrödinger was unwilling to retreat to normal science. In his later years he tried desperately to achieve a second breakthrough, a field theory that would unify gravitation and electromagnetism, but he was under no illusions about the likelihood of success. He told a young Irish scholar at Dublin: this problem is suitable only for old men who have already done something in science.

The Kuhnian analysis of scientific revolutions emphasizes the stress that precedes the breakthrough into a new world view. Schrödinger was a passionate man, a poetic man, and the fire of his genius was kindled by the intellectual tension arising from the desperate situation of the old quantum theory of Planck, Einstein, and Bohr. It seems that psychological stress, particularly that associated with intense love affairs, helped rather than hindered his scientific creativity. In his short autobiography, he wrote: "To create a true sketch of my life . . . is impossible, since leaving out relations with women in my case creates on the one hand a great void, and on the other hand seems to be required, firstly on account of scandal, secondly because they are hardly of sufficient interest, thirdly because in such matters no man is really completely sincere and truthful, nor ought he to be."

Erwin Schrödinger had the most complex personality of all the creators of modern physics. He was a passionate foe of injustice, yet he regarded any political activity as contemptible. He despised pomp and circumstance, yet he took a child-like pleasure in honors and medals. He was devoted to the vedantic concept that all men are members of one another, yet he eschewed co-operative work of any kind. His mind was dedicated to precise reasoning, yet his temperament was as volatile as a prima donna's. He claimed to be an atheist, but he always used religious symbolism and believed his scientific work was an approach to the godhead. In many respects he was a true son of Austria. As Robert Musil described in his *Der Mann Ohne Eigenschaften*: "In this country one acted – sometimes indeed to the extreme limits of passion and its consequences – differently from the way one thought, or one thought differently from the way one acted. Uninformed observers have mistaken this for charm, or even for a weakness in what they thought was the Austrian character."

Psychology teaches that during his life an individual tends to repeat patterns of interpersonal relationships learned in childhood; thus the form of living and even the sexual politics of a man are often determined by his life as a child and by the family in which he was brought up. As the only child in an indulgent family, Erwin might easily have come to believe that the world revolved around his center.

The social milieu of Schrödinger's formative years in Vienna must have influenced his philosophy of life, and through this the way in which he conceived his scientific work. It is difficult not to become enchanted by the intellectual and artistic brilliance of Vienna at the turn of the century, but the gleaming of Vienna was superficial, like the phosphorescence covering decay. While Schrödinger was attending the university and enjoying the delights of theater parties, new wine, and excursions to the nearby mountains, young Adolf Hitler, twice rejected as an art student at the Academy, having pawned his overcoat to buy bread and milk, was prowling about the snowy streets in his shirtsleeves, with uncut hair and a straggly beard, trying to sell a few pictures of Viennese churches. He told a friend, "I don't want to seem arrogant, but I think the world really lost something worthwhile when I was rejected by the Art School."

After all, the essence of the life of a creative person is

expressed in his work: the music of Mahler, the poetry of Hofmannstahl, the novels of Musil, and the physics of Schrödinger. For a theoretical physicist a causal chain between his scientific work and the structure of his personality is often not easy to discern. A great discovery may be to some extent accidental, the product of circumstance falling into a fertile field of intellect. Schrödinger touched upon this point in one of his poems:

> *Parabel*
> Was in unserem leben, freund,
> wichtig und bedeutend scheint,
> ob es tief zu boden drücke
> oder freue und beglücke,
> taten, wünsche und gedanken,
> glaube mir, nicht mehr bedeuten
> als des zeigers zufallschwanken
> im Versuch, den wir bereiten
> zu ergründen die natur:
> sind molekelstösse nur.
> Nicht des lichtflecks irres zittern
> lässt dich das gesetz erwittern.
> Nicht dein jubeln und erbeben
> ist der sinn von diesem leben.
> Erst der weltgeist, wenn er drangeht,
> mag aus tausenden versuchen
> schliesslich ein ergebnis buchen –
> Ob das freilich uns noch angeht?

Parable
My friend, what in this life / Weighty and important seems, / Whether causing dark depression / Or gladness and rejoicing, / Deeds, thoughts and wishes / Believe me, mean no more / Than a pointer's fluctuations / In an experiment that we design / To fathom Nature: / Merely molecular collisions. / Nor does the light spot's crazy flutter / Let you smell out the basic law. / It's not your joy and trembling / That makes sense of this life. / The World Spirit, if it goes about it / May from a thousand experiments / Enter finally a result – / Is it really any of our doing?

Scientific fame did not cause any marked change in the personality of Erwin Schrödinger, but it may have allowed him to express himself more freely. Max Born, who was shocked by some of the things he did, nevertheless wrote: "His private life seemed strange to bourgeois people like ourselves. But all this does not matter. He was a most lovable person, independent, amusing, temperamental, kind and generous, and he had a most perfect and efficient brain."

CHAPTER I

Family, childhood, and youth

Heredity or environment, nature or nurture, there is no general solution to the problem of how much each contributes to the structure of a personality or the achievements of a person. It has been said that any medical student can become an adequate surgeon provided he or she is willing to work long hours, follow the hospital rules, and placate the nursing staff. In other words, this particular skill is not significantly determined by genetic factors. No such prescription can be applied to mathematical physics; original work in this field requires a special kind of native aptitude, the genes must be right. This is a necessary but not a sufficient condition. A potential Schrödinger may be lost to physics if he happens to be born in a Bengali village rather than a university town.

Many theoretical physicists have come from academic families, which, at least in Germanic countries, are considered to be upper-middle class on the social scale. The fathers of Max Planck, Niels Bohr, Max Born, Wolfgang Pauli, and Werner Heisenberg were all university professors, of Law, Biology, Anatomy, Colloid Chemistry, and Byzantine Studies, respectively. Thus the ancestry and family history of Erwin Schrödinger are relevant to the story of his life, even if the relative contributions of genetics and economics cannot be clearly separated. Fortunately his aunt Rhoda compiled a history of his mother's side of the family; less is known about his father's side.

FAMILY HISTORY

Alexander Bauer, the maternal grandfather of Erwin Schrödinger, was born 1836 in Magyarovar in north-west Hungary. He studied mathematics and science at the University of Vienna, but transferred to the Polytechnic Institute (later the Technical University) to specialize in chemistry under Anton Schroetter, the discoverer of red phosphorus. Erwin's maternal grandmother was English. Her ancestry has been traced back to a Norman family, Forestière, whose stronghold was Bamborough Castle near Durham. The name was anglicized to Forster. Thomas, born in 1772, was the son of Colonel Forster, the governor of Portsmouth; he married Eliza Walker and they had five children. They lived in Kensington where their eldest daughter Ann was born in 1816. She was Erwin's great grandmother, whom he was to meet when he visited England as a child.

Ann married William Russell, a solicitor in Royal Leamington Spa, Warwickshire. He was descended from a family who had been concerned with the legal profession in Warwickshire for many years. The Russells had three children, William, Emily (who was called Minnie in the family), and Ann (called Fanny). Emily was born in Leamington Spa on September 14, 1841. She was baptized in All Saints Church, Leamington Priors (the old name of the Spa), which was the first parish of the Church of England in the town. The family lived in a spacious house with gardens at the back extending to the river Leam.

How did it happen that Emily Russell met and married Alexander Bauer of Vienna? The causal factor was that Emily's brother William was a chemist. He and Alexander became friends in Paris, where they were studying chemistry in 1859. Bauer was working at the *Ecole de Médecine* as a pupil of Charles Adolphe Wurtz, known to every beginning student of organic chemistry as the discoverer of the Wurtz synthesis of hydrocarbons, whereby two alkyl halides are linked by the agency of metallic sodium. Emily and her mother visited Paris on their way to a holiday at Montreux, and William took his friend Alex to meet them at their hotel. It was love at first sight between the serious young scientist and the lovely nineteen-year-old English girl. After this romantic meeting in Paris, Alexander had to return to Austria to establish himself in his profession. Two years later he was able to ask Emily to be his wife, and he was promptly accepted. He journeyed to Leamington Spa and on December 21, 1862, they were married in the parish church where Emily had been christened.

The newly-weds went to live in a charming small apartment at 20 Kärntnerstrasse, in the heart of old Vienna, not far from the house where Mozart had composed *The Marriage of Figaro*. Alexander registered his marriage at his parish church, the ancient and impressive cathedral of St. Stephan. In the summer of 1863, the Bauers visited Leamington Spa, where their first child was conceived. Alex returned to Vienna, while Emily stayed with her family until the spring of 1864, when she traveled to Vienna, arriving with little time to spare before the birth of the baby on Easter Sunday. This was Rhoda, who became Erwin's most doting aunt. A second daughter was born three years later, on March 27, 1867. This was Georgine (Georgie), the future mother of Erwin.

In 1866, Alexander lost an eye in an explosion while he was carrying out a chemical preparation. This accident lessened his interest in experimental research, and he turned increasingly to teaching, administration, and chemical history; he became famous as "the Nestor of Austrian chemistry." He was appointed to the chair of general chemistry at the Polytechnic, where he served till his retirement in 1904.

In 1874, Emily died of pneumonia shortly after the birth of her third daughter, Minnie. Alexander was desolated by the loss of his young wife, but after a year of mourning he revived and married seventeen-year-old Natalie Lechner. Natalie was a young woman of unusual independence and considerable literary ability; she was an excellent musician, playing the viola in a professional string quartet. The position of a young step-mother is always difficult, but it is likely that Natalie, with her strength of character, managed the rather docile Bauer girls without too many problems. She taught Georgie the elements of violin playing. After ten years, however, Alexander and Natalie agreed to separate, an event which at the time (1885) caused some negative comment. In 1890, Natalie met Gustav Mahler and for twelve years was his constant companion, until he married Alma Schindler in 1902.

After the departure of Natalie, Alexander devoted himself to his daughters and to a steadily increasing variety of professional and civic duties. He became a noted *Salonlöwe* (salon lion) in bourgeois Vienna society. In 1904 he was honored by being designated as *Hofrat*, which may be roughly translated as state councilor. In Imperial and Royal (K & K) Austria-Hungary, honorific titles were highly regarded by members of the upper-middle class, as indeed they still are today in modern Austria. Almost all university graduates were addressed as "Doctor" and with increasing distinction, the titles became increasingly ornate. Alexander once questioned some advice by his old professor *Hofrat* Schroetter, and was told quite seriously "Young man, a *Hofrat* never makes a mistake."

The three Bauer daughters all espoused men in technical professions whom they met through their father. On August 16, 1886, Georgie married Rudolf Schrödinger, who had studied at the Technical University, and who had inherited a small but profitable linoleum and oilcloth factory and wholesale business, and thus was acceptable to Bauer as a son-in-law. The marriage took place in the Lutheran Stadtkirche, and the registry indicates that the bridegroom was a Catholic and the bride Georgine Emilie Brenda Bauer was a "convert" to the Evangelical Church. One may suppose that the Bauer girls were all brought up in the Lutheran religion since it would be difficult to find Anglican churches and instruction in Austria. The rather tangled religious history of the Bauer family explains how Erwin Schrödinger came to be nominally a Protestant, despite his Catholic background in predominantly Catholic Austria.

Rudolf Schrödinger was born in Vienna on January 27, 1857, the son of Josef and Maria Bogner Schrödinger. He was baptized Rudolf Josef Carl in the parish church of St. Peter and St. Paul in Erdberg, an outer district of Vienna. Both the father's and mother's families had lived in Vienna for three or four generations, but the Schrödingers originally came from the Oberpfalz area in Bavaria. Rudolf's mother, Maria Anna Josepha Bogner, was a nineteen-year-old orphan when she was married in the Catholic church of St. Carl on May 14, 1853. Her father had been the proprietor of a coffee house in the suburbs. She had three children, a son Erwin, who died as a child, a daughter Marie, and their last child Rudolf. Rudolf's mother died at the age of twenty-four, six days after the stillbirth of a fourth child. Rudolf was less than two years old when his mother died; his bereaved father did not remarry and undertook the upbringing of his two children.

About three months after their marriage, Rudolf and Georgine effected that fortuitous combination of genes that produces an individual of genius, and on August 12, 1887, Erwin was born. He was born at home, Apostelgasse 15, in Erdberg, Vienna 3. Georgie, who loved the works of Goethe, wanted to name him Wolfgang, but Rudolf, not usually given to sentimental considerations, favored the name Erwin, after his long-departed elder brother. The circumstances of the baptism of the child were unusual. It took place at the home of his grandfather Alexander Bauer, Kärntnerstrasse 20. On October 17, the minister of the Evangelical Church came to the house and baptized the baby Erwin Rudolf Josef Alexander. The godfather was Alexander Bauer and there was no godmother. Such a home christening was not unusual but the parental home would have been the expected place. A possible scenario is that Rudolf, who was not a practicing Catholic, did not wish to have the baby baptized at all, but that Alexander decided to have it done anyway, and out of deference to his daughter's religion, asked the Evangelical pastor to come to his house and administer the sacrament.

Whether through choice or necessity, Rudolf and Georgie had no more children. As an only child, Erwin received the full attentions of his mother, and for many years hardly less from Aunts Rhoda and Minnie, as well as the services of a succession of young maids and nurses, all of whom considered Erwin to be a budding genius deserving constant adulation. Raised in such an atmosphere of tender, loving, feminine care, it is hardly a wonder that Erwin grew to depend upon it and to expect it as his due, such expectations being fully realized throughout his lifetime.

CHILDHOOD

In 1884 a splendid town house, five storeys high, was constructed at 3 Gluckgasse, a fashionable street just off the Neumarkt in the center of the First District of Vienna. The house was purchased by Alexander Bauer not long after its completion, and divided into spacious apartments, only one on each floor. In 1890 he rented the top-floor apartment to the Schrödingers. This was to be their Vienna home until the year before the death of Erwin's mother in 1921. The front windows looked over tiled roofs to the serrated Gothic spire and flying buttresses of St. Stephan's Cathedral.

Life in the later years of the Habsburgs was pleasant and unhurried for those with ample incomes. Political freedom was not essential to this good life and its absence did not trouble the ordinary citizen, although there were grumblings from subject peoples in distant parts of the Empire. In the words of the socialist Viktor Adler, the government was "a despotism mitigated by slovenliness." If anything went wrong, there were always Jewish plots to blame, and the endemic anti-semitism helped to satisfy both social and religious needs.

Erwin's childhood was somewhat different from that in a typical well-to-do Vienna family owing to the absence of siblings and the prevalence of aunts. Some recollections have been given by his Aunt Minnie. She was fourteen years older than Erwin but they were great companions. He learned from Minnie to speak English before he even spoke German properly. She brought him a book of Bible stories from England; he was not terribly impressed by the stories but they were his earliest reading in English. Even before he could read or write, he kept a record of day-to-day happenings, which he dictated to Minnie. He maintained this habit of keeping a diary all his life. The later books, called *Ephemeridae*, contain everything

1. Erwin and Grandfather Alexander Bauer (c. 1890)

except scientific material. He must have thought that science had a certain permanence, while all else was ephemeral. From one of his earliest records, in 1891 he describes how: "In evening Aunt Emmy cooked a good supper and then we spoke all about the world." Minnie recalls that: "Despite my small knowledge of the subject, he was especially interested in asking me about astronomy. For example, I would stand and represent the earth and he would be the moon and run around me, and both of us would move slowly about a light that represented the sun."

We know, however, from Erwin's own testimony that the influence of his father was always most important:

To my father I am thankful for far more than only this, that he gave us a very comfortable life, and assured for me an excellent upbringing and a carefree university education, while till almost the end of his life he carried on with little zeal or talent the inherited, prosperous oilcloth business. He had an unusually broad culture; after his university studies as a chemist, there followed many years of intense concern for Italian painters, accompanied by his own landscape drawings and etchings, giving way finally to pyxides and microscope, from which arose a series of publications on plant phylogenetics. To his growing son, he was a friend, teacher and inexhaustible conversation partner, the court of appeal for everything that might sincerely interest him. – My mother was very good, cheerful by nature, sickly, helpless in the face of life, but also undemanding. Besides sacrificial care, I have her to thank, I believe, for my regard for women.

Georgie had an interest in playing the violin but after it found no response from her husband, her art wasted away. Almost uniquely among theoretical physicists, not only did Erwin not play any instrument himself, he even displayed an active dislike for most kinds of music, except the occasional love song. He once ascribed this antipathy to the fact that his mother died from a cancer of the breast, which he thought was caused by mechanical trauma from her violin. More likely he learned this distaste for music as a child, echoing his father's lack of response to his mother's art.

Erwin showed early signs of a brilliant mind and his father was anxious lest he be pushed too quickly. He was not sent to the elementary school, but received lessons at home two

2. Erwin with Aunt Emily (Minnie) Bauer (c. 1893)

mornings a week from a private tutor. This arrangement was not unusual in upper-middle-class families. Their principal goal was to ensure that their boys passed the entrance examination for the Gymnasium, usually taken when they were nine or ten years old.

Instead of taking this examination at the usual time, Erwin had a long holiday and was taken to England by his mother and Minnie to visit the Russells at Leamington Spa. This was in the spring of 1898 when he was ten. He saw his great grandmother at her home on Russell Terrace, named after his great-great grandfather. He explored Kenilworth and Warwick Castles, he rode the donkeys on the wide beach at Ramsgate, and also learned to ride a bicycle, a skill he was to practice all his life for both pleasure and utility.

On the trip back to Austria, they crossed the channel from Dover to Ostend, and then visited the beautiful medieval city of Bruges. Proceeding to Cologne, they boarded the Rhine steamer and traveled through the country of wine and castles via Koblenz and Rüdesheim to Frankfurt-on-Main, where they took the train to Munich and thence to Innsbruck.

Here Erwin had his first experience of school. His parents were worried about the entrance examination and "afraid that I might have forgotten my ABC's." He went to the St. Niklaus School for a few weeks. His mother tried to preserve an English atmosphere, by insisting when they walked in the park: "Now we are going to speak only English to each other." She also tried to observe the English custom of half-holidays, but amid all his other holidays these may have been difficult to distinguish. Throughout his life Erwin seldom allowed anything to interfere with his vacations and holidays.

AKADEMISCHES GYMNASIUM

As was to be expected, he passed the examination easily, and entered the Akademisches Gymnasium in the autumn of 1898, having just turned eleven. He was about a year older than most students in his class. He had never experienced the rough and tumble socialization of the schoolroom and playground, nor any competition with siblings at home. He had been raised almost exclusively in the company of adults, and it was perhaps only to be expected that he would retain something of the *enfant terrible* when he reached manhood.

His Gymnasium was the most secular, the least religiously oriented, of all those in Vienna. Ludwig Boltzmann, Arthur Schnitzler, and Stefan Zweig had formerly been pupils there. It is situated on Beethoven Platz, just off the Schubert Ring, about ten minutes walk from Erwin's house on Gluckgasse. Djuna Barnes once pictured the schoolboys of Vienna as "flocks of quail, taking their recess in different spots in the sun, rosy-cheeked, bright eyed, with damp rosy mouths, smelling of herd childhood, facts of history shimmering in their minds like sunlight, soon to be lost, soon to be forgotten, degraded into proof."

As summarized by Friedrich Paulsen, a contemporary authority, the ideal of education in the Gymnasium was to graduate

a human being whose faculties enable him to form a clear and definite conception of the actual world, and who, by virtue of his will, is able to follow his original bent, whose imagination and fine emotions are trained to the perception of the beautiful and the heroic. This is a man in the full sense of the word; this is true humanistic culture.

The education designed to achieve this admirable result was based on an intensive study of the Greek and Roman classics. Since virtually the only way to enter the university was through the Gymnasium, all university students began with a shared background in humanistic studies. Such a preparation was welcomed by many who later became eminent scientists. Max Laue, for example, wrote:

I doubt that I should ever have devoted myself entirely to pure science if I had not at that time come into that inner harmony with Greek language and culture, which the humanistic Gymnasium and no other kind of school provided . . . If you wish to bring out scientific development later, I have a recipe: send the youth to a Gymnasium and let him learn the ancient languages.

Of course, there is no single recipe for the making of scientists: most American physicists have been illiterate in ancient

3. Erwin with his parents Rudolf and Georgine on holiday in Kitzbühel

tongues, yet the best of them have also done great physics. Besides the Latin and Greek classics, there were classes each year in German language and literature with emphasis on Goethe and Schiller; and of course in Vienna considerable attention to the great Austrian dramatist Grillparzer. Only three hours a week were spared for mathematics and science. Mathematics consisted of algebra and geometry; they did not get so far as calculus.

Erwin was reasonably happy with the Gymnasium. "I was a good student, in all subjects, loved Mathematics and Physics, but also the strict logic of the ancient grammars, hated only the memorization of incidental dates and facts. Of the German poets, I loved especially the dramatists, but hated the pedantic dissection of their works." The schoolwork was easy for him and he had ample time for other things.

From entry till graduation in 1906, Erwin was always first in his class. Later a schoolmate wrote:

I can't recall a single instance in which our Primus ever could not answer a question. Thus we all knew that he took in everything during the instruction, understood everything, he was not a grind or a swot. Especially in physics and mathematics, Schrödinger had a gift for understanding that allowed him, without any homework, immediately and directly to comprehend all the material during the class hours and to apply it. After the lecture of our professor Neumann – who taught both subjects during the last three gymnasium years – it was possible for him to call Schrödinger immediately to the blackboard and to set him problems, which he solved with playful facility. For us average students, mathematics and physics were frightful subjects, but they were his preferred fields of knowledge.

In fact he did not dispense with homework. If anyone in the house asked where he was, the answer was always "He is upstairs in his room and is studying." Erwin had two small rooms of his own overlooking the courtyard at the back of the house.

On two afternoons a week, Erwin returned to school for instruction in the Lutheran religion. "From this I learned many things, but not religion." He had also received the required religious instruction from his tutor as a preparation for the Gymnasium. His favorite question at the end of a bible story was "Herr Teacher, do you really believe that?" Thus it was not for lack of formal catechism and bible studies that Erwin became indifferent, sometimes even inimical, to organized religious beliefs and practices. His mother was mildly religious, but he learned his negative attitude from his father, and then reinforced it through his own reading, observation, and experience. Among the many anecdotes about Erwin's boyhood, Aunt Minnie recalled not once dealing with a religious observance, not even at Christmas. The Schrödingers never entered a church except to be married or to be buried, or to attend similar rites for their friends. Nevertheless, Erwin had a great respect for saints and mystics. He became anti-clerical rather than basically anti-religious. It was the history of the Church rather than its beliefs that distressed him.

One subject not included in the curriculum, except for its cursory denunciation as a heresy in the religion class, was the Darwinian theory of evolution. Erwin was able to go into this

4. Schrödinger as Gymnasium student

subject at considerable depth during walks and botanical
excursions with his father.

On the basis of his botany, my father advised caution. The melding of
natural selection and the survival of the fittest with the De Vries
mutation theory had not yet been made. I don't know why zoologists
were more ardent Darwinians than botanists. My father's friend,

Hofrat Anton Handlirsch, a zoologist at the Natural History Museum, taught that development is causal not "final," there is no entelechy.

Erwin was soon converted to this view. "Naturally I was an enthusiastic Darwinian, which I still am today." [1960]

His best friend in high school was Tonio Rella, who later became professor of mathematics at the Technical University in Vienna. Tonio was always second and Erwin first in class during all their eight years at the Gymnasium. On the last day of the war, in April, 1945, Rella was killed by a stray Russian shell, so that the reunion Schrödinger had anticipated when he returned to Austria never took place. The Rella family owned a country inn, Kastell Küb, in Semmering, and Erwin often used to spend holidays with them there. The inn was surrounded by extensive wooded grounds with fine views of the Schneeberg. Both boys were ardent hikers and mountain climbers. Grandfather Bauer had been one of the discoverers of the beauties of Semmering and one of his most loved pursuits was the ascent of Sonnenwendstein, the mountain that shelters the town. Erwin and Tonio often followed the trails that he had blazed forty years earlier.

In the course of time, there came to be another attraction at Semmering, more powerful even than the lure of the mountains. This was Tonio's sister Lotte, called "Weibi," with whom, Erwin later said, "I was fairly permanently in love" (*ziemlich dauerhaft verliebt*). Weibi was an Italianate beauty with dark brown eyes and a generous figure; in an earlier time, she might have been a model for a Caravaggio portrait.

Erwin almost always used to spend his Christmas holiday from about December 22 till the new year at Semmering with the Rella family, but for Christmas Eve he would have to return to Vienna to be with his own family, which included his grandfather and aunts. He was unhappy to leave his sweetheart, and anyone who can recall an adolescent first love will understand exactly how he felt, a reluctant member of the family circle whose thoughts were wandering elsewhere. In those days, such a teenage romance would be limited to worship from afar, holding the beloved for a Vienna waltz, perhaps a gentle kiss if rare opportunity occurred, but the sister of a school friend would be inviolable and a more intimate relationship only a subject such as dreams are made of. Lotte's parents would have been happy to encourage a match with Erwin. His devotion to her was more than a mere *Kinderliebe*, and it would leave an indelible impression on his soul, but he was not prepared at this time to consider any enduring emotional ties.

A WORLD OF THEATER

For the students, the years at the Gymnasium spanned the onset of puberty and the pangs of adolescence. They entered as ten-year-old boys, and left as sexually mature young men of eighteen. Sexuality was an anarchic factor and bourgeois society protected itself by establishing a dual system of behavior. School, family, popular literature, and newspapers formed a world in which sex hardly existed outside the privacy of the conjugal bedroom. Parallel with this conventional world of bourgeois morality, was a flourishing underworld of pornography and prostitution. Sex for pay was not, however, confined to casual pickups and brothels. For the wealthy, there was a *demi-monde* of music halls and artists' studios where attractive mistresses awaited those able to support them, but such pleasures were outside the financial range of most students. There was, however, a charming institution known as *Die süsse Wiener Mädel*, perhaps a shopgirl whose greatest ambition was a sacrificial affair with a university student. Arthur Schnitzler was particularly expert on the subject of the "sweet Vienna maid."

The Vienna theater, at one of the high points in its history, provided Erwin with at least vicarious experience in the world of romance, through the works of the great dramatic poets at the Hof-Burg Theater on the Ringstrasse. Josef Kainz, believed by many critics to be the greatest actor of all time, was creating some of his most famous roles, Hamlet, Cyrano, Tasso, Orestes, Mephisto, Don Carlos. The emperor's favorite, Katherina Schratt, was one of the leading ladies. Vienna audiences and critics were the most knowledgeable in the world; no *Schlam-perei* (sloppy work) was tolerated. Special matinées were held on Sunday afternoons for students and workers' alliances. As Stefan Zweig remarked, this theater was more than a mere stage for plays, it was a microcosm in which Austrian society could see itself.

Erwin loved the theater, going as often as he could, some-times twice a week, and he kept an annotated record of the performances. As a student, his favorite playwright was Franz Serafikus Grillparzer, who was born in Vienna in 1791, the year Mozart died. A shy man, torn all his life between the demands of his work and romantic loves for beautiful but inaccessible women, he was a master of erotic drama. An example from his *Sappho*:

> He who knows what love, what life is, man and woman,
> Does not weigh man's love 'gainst woman's passion.
> Most fickle is man's fitful disposition,
> Subservient to life, most fickle life.

These were stirring dramas for an adolescent student. They must have helped to convince Erwin that there is more to life than books and theory.

Some pages from Erwin's *Theater Notizbuch* have been pre-served, dating from the autumn and winter of 1904–5, his next to final year in high school. The first notice is of one of the greatest dramas in German literature, Schiller's *Wallenstein*, which he found most impressive with Sonnenthal in the prin-cipal role at the Burgtheater. On September 3, 1904, the play at the Raimund Theater was *Herodes and Mariamne*, a tragedy in five acts of blank verse by Friedrich Hebbel (1813–63). King Herod was played by Wiecke, and Erwin thought that he surpassed even Kainz, having more temperament and more warmth, "a thin figure of unbelievable elasticity." On October 8, Erwin got a seat in the upper-most gallery, "with the perspective of a bird," to see Hebbel's *Gyges and His Ring*. Erwin commented that

I do not think that a young person of my age can understand this play, not so much the external plot, but just as one can imagine the emotional processes of others when one has experienced them even once in any form, so also with the stage one can enter into an emotional feeling, if one has even once felt something similar, and that obviously does not hold true for the emotional feelings of Rhodope.

On January 26, 1905, he saw Schiller's *Don Carlos*, a marathon production, which began at 6:30 p.m. and ended at 12:30 a.m. The title role was played by Kainz, not entirely to Erwin's satisfaction: "I have read Schiller's *Letters about Don Carlos* with great interest and believe that I shall gradually come to a complete understanding of the play."

   The art world of Vienna at the turn of the century was even more erotic than the avant-garde theater, since it had less need to appeal to a paying bourgeois audience. In 1897, Gustav Klimt led a revolt of young artists in founding a movement known as the *Secession*. In 1894, Klimt, then a brilliant but fairly conventional artist, received a commission to paint three heroic ceiling paintings for the great hall of the new university, to represent philosophy, medicine, and jurisprudence. By the time the first two paintings were completed, in 1900 and 1901, he had fervently embraced the ideas of Wagner, Nietzsche, and Schopenhauer, translating them into powerfully symbolic pictures, filled with disturbingly erotic female figures who seemed to drown the masculine arts of philosophy and medicine in a sea of hopeless sexuality. This was not what either the university professors or the liberal government had anticipated, and a storm of public protest and personal abuse broke over the unrepentant artist. He was accused by the gutter press of being a mere pornographer, by the clericals of being in league with seditious Jewish philosophers, by the academics of being anti-intellectual, and the social democrats bemoaned his lack of optimism for the future of society. The controversy raged for some years; the government tried to stand firm but the paintings had to be removed to the Modern Art Museum and never adorned the university.

   How was Erwin affected as an adolescent by the erotic art of avant-garde Vienna? Among the papers that he kept all his life were copies of the journal of the University Science Club. The issue for 1909 has an erotic drawing more or less copied from Klimt's "Fish Blood." Like so many of Klimt's drawings and paintings, it is a desperate effort "to capture the feeling of femaleness." Much of Erwin's erotic life would be devoted to a similar effort, not to master or dominate women, but to capture the essence of their sensuality by experiencing it with them and through them.

All this was for the future. The avant garde was certainly not the world of Erwin Schrödinger and Tonio Rella when they entered the university together in the fall of 1906. Erwin's romance with Lotte Rella continued, but it remained within the bounds set by conventional morality. We know this from his *Ephemeridae*, where he carefully recorded the names of all his loves with a notation to indicate the denouement.

CHAPTER 2

University of Vienna

Vienna University is the second-oldest German-speaking university, founded in 1365 by Pope Urban V, seventeen years after the University of Prague. Albert of Saxony, formerly rector of the University of Paris, brought the hitherto secret statutes of that establishment to Vienna as a basis for the new university, and he became its first rector. Students came from all over Europe, but especially from Austria, Bohemia, Saxony, and Hungary.

The first professorship of physics (natural philosophy) was established in 1554, but there was no equipment for experimental work till 1715 when a collection of apparatus was made by the Jesuits. Andreas Baumgartner, who became professor in 1823, was the first to give physics lectures in German instead of Latin. He helped to found the Vienna Academy of Sciences and served as its president for fourteen years. Through his efforts, the teaching of physics in Austria was modernized, and brought up to the standards of Germany, France, and Italy.

In the spring of 1848, popular revolutions swept through Europe. In Vienna, workers led by students of the "Academic Legion" marched through the streets with red banners, hammers, and scythes. A unit of the Viennese Grenadiers joined the rebels and the entire city was soon in their hands. By autumn, Vienna had been recaptured and the Army took revenge for its ignominious defeat by the "intelligentsia in arms." It seized the university and dispersed the staff into makeshift buildings in outer districts. Some concession to the prevailing discontent seemed to be required, however, and the half-witted Emperor Ferdinand was forced to abdicate in favor of his eighteen-year-old nephew Franz Joseph.

UNIVERSITY PHYSICS

In 1850, Christian Doppler, a native of Salzburg, was appointed to the chair of physics. He was famous for his discovery that if a source of light is in motion relative to an observer, the observed frequency is altered. The analogous effect on sound waves is familiar to anyone who has ever listened to the mournful notes of a passing locomotive. Schrödinger, who was always interested in anything related to the Vienna school of physics, published a paper on the Doppler effect many years later. When Doppler retired in 1853, Andreas Ettinghausen, who had served from 1835 to 1848, was reappointed. He supervised the move of the Institute of Physics to its new quarters, an inadequate building at Erdbergstrasse 15, in the Third District of Vienna, not far from where Erwin was born.

In 1884, most of the university moved to a magnificent new building on the Ringstrasse, but the physicists remained exiled in Erdberg, although by 1875 their laboratories had been moved to a ramshackle temporary building on Türkenstrasse closer to the new university site. The successor of Ettinghausen to the chair of physics was Josef Stefan. His most famous discovery was concerned with the properties of what was then called "black radiation." All bodies are continuously absorbing and emitting radiation. When a body is in equilibrium with its environment, the radiation it is emitting must be equivalent in wavelength and energy to the radiation it is absorbing. It is possible to imagine a body that is a perfect absorber of all the radiation incident upon it, an "ideal black body." Stefan, in 1879, discovered experimentally that the energy of radiation from such a body depends upon the fourth power of the absolute temperature. A few years later, a former student of Stefan's, Ludwig Boltzmann, derived this law theoretically, and it became one of the corner stones of radiation physics. The interpretation of the wavelength distribution of black radiation by Max Planck led directly to the quantum theory.

Another important scientist at the university at this time was Josef Loschmidt. While an assistant to Stefan, he made a major contribution to science, one that required neither experimental nor mathematical skill, but simply the ability to see an important relation that others had missed. As a result of the work of James Clerk Maxwell and Rudolf Clausius, the kinetic molecular theory of gases was becoming generally accepted: a gas consists mostly of empty space in which tiny molecules are flying about at great speeds, colliding with one another and with the walls of their container. The average effect of the collisions with the walls is the pressure exerted by the gas. A difficulty with this model was that nobody knew either how many molecules are in a given container of gas or the size of the molecules of any particular gas. Loschmidt made the necessary calculations of these quantities and reported the results at a meeting of the Vienna Academy. This was the most exciting thing that had ever happened to molecules – for many physicists, they ceased being hypotheses and became realities.

A scientist can have only one biological father, but he can have several scientific fathers, whose influence on his life may be even more important. Thus this history of physics in Vienna traces the scientific ancestors of Schrödinger. He had two scientific fathers, Franz S. Exner in experimental physics and Friedrich (Fritz) Hasenöhrl in theoretical physics. Loschmidt and Stefan were scientific grandfathers of Schrödinger on the Exner side.

Exner was appointed professor in 1891 when Loschmidt retired. At that time he had offers from Graz, Innsbruck, and Vienna, but he chose to stay in Vienna because he was looking forward to building a new Institute of Physics to replace the inadequate facilities at Erdberg, a hope that was not to be realized for another twenty years. Exner worked on many different physical problems, and he inspired his students with an enthusiasm for these problems, so that the Exner Circle, as they came to be called, shared not only a common university background but also common interests in their research fields.

5. Professor Franz Exner

These included electrochemistry, atmospheric electricity, radioactivity, crystal physics, spectroscopy, and the science of color (*Farbenlehre*). Schrödinger was to work in all the Exner fields except electrochemistry, but usually on theoretical problems, where the only equipment needed was pencil and paper.

In 1898 the Vienna Academy, on the recommendation of Exner, arranged for the gift to Marie and Pierre Curie of 100 kilograms of pitchblende residues from the St. Joachimstal uranium mines. From this material, the Curies achieved the first isolation of radium, and they sent Exner a small sample of highly enriched material, with which the Austrian scientists accomplished some noteworthy researches. The Academy then bought 10,000 kilograms of residues from St. Joachim, which were processed to yield four grams of pure radium chloride. In 1907 the Academy sent Rutherford and Ramsay in Manchester 400 milligrams of radium bromide, and their great discoveries were all made with Viennese radium. In 1908, an Institute for Radium Research was established in Vienna and many of the physicists worked on projects in radioactivity and related fields.

LUDWIG BOLTZMANN

We have described the scientific ancestry of Erwin Schrödinger on the experimental side, and now must consider his even more important scientific forebears on the theoretical side. Here the father was Friedrich (Fritz) Hasenöhrl and the grandfather was one of the greatest theoretical physicists of all time, Ludwig Boltzmann. A scientific family tree is not subject to biological constraints; Hasenöhrl was not only Schrödinger's father but also his elder brother, since he took his first degree under the supervision of Exner, in 1896.

Boltzmann was born in Vienna on February 20, 1844, the son of a government official. The date was Shrove Tuesday, and half seriously he used to ascribe the sudden changes in his spirit between happiness and depression to the fact that he was born during the dying hours of a gay Mardi Gras ball. He was short and stout with curly hair, his fiancée used to call him her "sweet fat darling." He received his doctorate from the University of Vienna in 1866 for work on the kinetic theory of gases under Stefan. His genius was soon recognized and at the early age of twenty-five he became professor of mathematical physics at the University of Graz.

Boltzmann's temperament was such that he never stayed very long in one place, but wherever he was, he continued to make major advances in the kinetic theory of gases. Quite independently of each other, Boltzmann and Willard Gibbs of Yale created statistical mechanics. This is the science that forms the connecting link between the small-scale world of atoms and molecules and the large-scale world of gases, liquids, and solids. The laws of the large-scale world, for example, the dependence of the pressure of a gas on its volume and temperature, can be derived as the statistical consequence of the mechanical behavior of the enormous number of molecules in any macroscopic volume. So far as the large-scale world is concerned, the positions and velocities of individual gas molecules are *hidden variables*, which underlie the ordinary laws of gas behavior but do not explicitly appear in the gas laws. A considerable part of Schrödinger's research was devoted to statistical mechanics, his publications in this field filling the 514 pages of Volume 1 of his *Collected Works*.

In Boltzmann's time, chemists and physicists were much concerned with the question of the "reality" of atoms and molecules. This concern was not the old philosophic conflict between "realism" and "idealism," although this too was mixed up in the scientific question. Many scientists, however, believed that diamonds and carbon atoms are entities of essentially different kinds, so that diamonds "really do exist" whereas carbon atoms are merely theoretical concepts used to derive mathematical equations. For Boltzmann, however, atoms were every bit as real as diamonds.

When Stefan died in 1894, Boltzmann was appointed to his chair at Vienna University. The next year Ernst Mach was appointed to the chair of history and philosophy of science. Mach was the most important opponent of the reality of atoms and his lectures and books enjoyed an enthusiastic following in

Vienna. In 1900, Boltzmann accepted a chair of theoretical physics at Leipzig. In 1901, however, Mach retired because of ill health and Boltzmann returned to Vienna, since his former chair there had not been filled. Rather ironically, considering his dim view of most philosophers, he was also assigned the course in philosophy that had been given by Mach. He was a realist and a materialist in philosophy, but in personal relationships he was very soft-hearted, he never failed a student. An ardent supporter of Darwinism, he saw in evolution the mechanism by which inanimate atoms had evolved into mechanical structures called human brains, capable of love, pity, and artistic creativity. Modern molecular biologists like Francis Crick and Jacques Monod would have felt perfectly at home with Ludwig Boltzmann.

As the attacks on atomism continued, the followers of Mach called Boltzmann "the last pillar of that bold edifice of thought." His health declined and he became more depressed, feeling each tremor of what he began to believe was a tottering edifice that would collapse with all his life's work under the rubble. In the summer of 1906 he went for a holiday to the beautiful Bay of Duino near Trieste and committed suicide by hanging himself, while his wife and daughter were enjoying a swim.

Although it was sometimes said that suicide was a way of life in Vienna, the loss of Boltzmann was a terrible shock to the members of the physics department. Erwin Schrödinger was personally broken hearted, for he had excepted to begin within a few months his studies in theoretical physics under the great master. He recalled his feelings of that autumn, when he entered the physics building:

The old Vienna Institute, from which shortly before Ludwig Boltzmann had been torn away in a tragic fashion, the building where Fritz Hasenöhrl and Franz Exner worked and many another of Boltzmann's pupils went in and out, engendered in me a direct empathy for the ideas of that powerful spirit. For me his range of ideas played the role of a scientific young love, and no other has ever again held me so spellbound.

FRITZ HASENÖHRL

Lectures in theoretical physics at the University were suspended for eighteen months, until Hasenöhrl was appointed to the vacant professorship. Like most of his colleagues he was from the comfortable middle class. He had graduated from the University in 1896. In 1898, Kammerlingh-Onnes, founder of the famous low-temperature laboratory at Leyden, had asked Boltzmann to recommend a research assistant for a year, and Boltzmann sent him Hasenöhrl as his "best student." At Leyden he also came to know the great Dutch theoretician Hendrik Lorentz, who has been called the spiritual link between Maxwell and Einstein. He returned to Vienna in 1899 as *Privat Dozent* and in 1904 published his most notable paper, which in some ways anticipated the famous equivalence of mass and energy,

$$E = c^2 m,$$

derived from Einstein's special theory of relativity (1905). Hasenöhrl showed that a moving charged particle has an inertial mass that depends upon its velocity and that a volume of black radiation has a mass proportional to its energy. Unfortunately he did not get the correct proportionality factor c^2, otherwise he would have achieved greater renown. He was thirty-three years old when he succeeded Boltzmann as professor of theoretical physics in 1907.

His inaugural lecture was a masterly synthesis of the statistical theories of Boltzmann and an exposition of the philosophy of the great scientist. Erwin listened with wonder and excitement. He resolved to make mathematical physics his life's work, following in the footsteps of these great Austrian masters.

ERNST MACH

Although Boltzmann and Mach had engaged in a hard-fought battle over the reality of atoms, most of the physicists seemed to have little difficulty in accepting the philosophical ideas of the

6. Professor Fritz Hasenöhrl in 1915

latter about the basic nature of science, while using the methods of the former in their daily work. As Schrödinger explained:

Filled with a great admiration of the candid and incorruptible struggle for truth in both of them, we did not consider them irreconcilable. Boltzmann's ideal consisted in forming absolutely clear, almost naively clear and detailed "pictures" – mainly in order to be quite sure of avoiding contradictory assumptions. Mach's ideal was the cautious synthesis of observational facts, which could, if desired, be traced back to the plain, crude sensual perception . . . However, we decided for ourselves that these were just different methods of attack, and we were quite permitted to follow one or the other provided we did not lose sight of the important principles . . . of the other one.

The most important scientific research of Schrödinger in the period 1919–24 was in the field of color theory, and the subject matter and methodology appear to have been inspired by Machian principles.

What was Mach's view of the world? The first basic component of a philosophical view is epistemology (*Erkenntnistheorie*), what and how does one know? The second basic philosophical component is ontology, what exists? Although Mach's views developed and changed somewhat in the course of his life, his mature position was that all we can know of the world are what he called *elements*. The elements include sensory perceptions but in addition other things that can be reliably measured, such as space and time. Elements that refer to the human body are called *sensations*. The world is directly presented to us in terms of these elements. Thus the epistemology of Mach is called *presentational phenomenalism*. It is important to distinguish this from *representational phenomenalism*, which states that the external world cannot be directly sensed, but is represented to consciousness, so that all appearances are actually mental, but a real external world can be reliably known through inference from empirical evidence of the senses.

Mach was not only a presentational phenomenalist in his epistemology, but also in his ontology. A world consisting of the elements is what actually exists; there is no other "real world" hiding behind the phenomena, nor is there any ego or self hiding in the consciousness of human beings. The ego is simply a group of sensations like an apple or an iceberg. Mach believed that the aim of science is the ordering of the elements in the most economical way:

The world consists of colors, tones, warmths, pressures, spaces, times, etc., which now we do not want to call "sensations" or "phenomena," because in both names there lies a one-sided arbitrary theory. We call them simply "elements." The comprehension of the flux of these elements, whether directly or indirectly, is the actual aim of science . . . That the world in this sense is our sensation is not subject to doubt.

Mach recognized the utility of theories in suggesting connections between phenomena, but he did not think that they had any permanent value: "Theories are like dry leaves which fall away when they have long ceased to be the lungs of the tree of science." In particular he distrusted the idea that atoms are anything but convenient symbols for summarizing experience. "Certainly one must wonder how colors and tones, which are so close to us, can suddenly appear in a physical world of atoms, how we can suddenly be astonished that what outside so dryly knocks and clatters, inside the head can sing and shine. How, we ask, can matter feel, which also means how can a thought symbol for a group of sensations become itself a sensation?"

After he retired, Mach became especially interested in the theories of gravitation and electromagnetism, since they both involve fields that vary inversely as the square of the distance from a source. He thought that it might be possible to devise a unified field theory that would supplant current atomic theories. In 1909, he was attracted to the recent work of Einstein and to a book by Paul Gerber, *Gravitation and Electricity*. He asked Professor Wirtinger to find out from some of the theoretical physicists what they thought of the researches of Einstein and Gerber. Wirtinger passed this request along to Schrödinger and wrote to Mach on July 28, 1910. "I have given Gerber's paper to a young electron man, who otherwise seems quite reasonable, and he offered to give me a detailed opinion in return. Dr. E. Schrödinger has now written that detailed letter, and what seems striking to me is his objection that the whole thing [the relation between gravity and electromagnetism] is quite different when another kind of radioactive material is taken into serious consideration." We may never know what Schrödinger meant by this mysterious observation. He found parts of Gerber's treatise very obscure: "Regrettably, I could not carry out my task in a very satisfactory way, since some of the essential points in Gerber's paper remain completely unclear." Schrödinger and Einstein were both to devote many of their mature years to a fruitless search for the unity of electromagnetism and gravitation.

The ideas of Mach have had a great influence on the development of the philosophy of science in general, and on the interpretation of quantum mechanics in particular. From 1920 to 1950, logical positivism, which was formulated by the "Vienna Circle," originally the "Mach Circle," reigned almost unchallenged as accepted doctrine in scientific philosophy. Yet there are a number of fairly obvious defects in presentational phenomenalism. For instance, it fails to explain the close relationship between mathematical reasoning and theoretical physics: mathematical operations and symbols do not denote empirical sensations, and yet one cannot do science without them. Also, experiments are *interactions* of the scientist with the environment; how can they be explained as mere collections of sensations? Mach fails to explain the enormous predictive power of physical theories: how can it be that Dirac predicts a positive electron and Anderson finds it in a cloud chamber? Einstein and Planck, who were disciples of Mach in their younger days, eventually cast him aside. Yet Einstein wrote in his obituary of Mach: "Even those who think of themselves as Mach's opponents, hardly know how much of Mach's views they have, as it were, imbibed with their mother's milk."

As a young man, Schrödinger was able to maintain a pragmatic balance between Machian positivism and Boltzmannian realism, but as he grew older this compromise no longer satisfied him. In 1925, he wrote: "One recalls in memory the feeling of anxious, heart-constricting solitude and emptiness that I dare say has crept over everyone on first comprehending the description given by Kirchhoff and Mach of the task of physics (or science in general): a description of facts that is as far as possible complete and as far as possible economical of thought." Schrödinger was not willing to confine the beauty of science within the cold prison walls constructed by Mach and his followers.

THE UNIVERSITY STUDENT

When Erwin entered the university in the fall of 1906, he brought with him his reputation from the Gymnasium as an outstanding student, in fact something of a genius, and this reputation was soon confirmed by his brilliant performances in mathematics and physics. Hans Thirring, who entered in 1907, recalls his first encounter with Erwin in the library of the mathematics department. A student entered, steely gray-blue eyes and a shock of blond hair. Another student nudged Hans and said *Das ist der Schrödinger*. All the students regarded him as something special, but he was not cold and aloof, he often helped them with difficulties in maths or physics. He became a good friend of Hans who commented: "We saw in him a fire spirit at work, always breaking through to something original in every research."

His closest friend at the university, indeed he says the only really close friend he ever had, was not one of the physics students, but a student of botany, Franz Frimmel, whom he called Fränzel. They used to spend hours at a time talking together about philosophical questions, which Erwin at the time thought were quite original, but later realized were those exciting the mind of every adolescent student. Sometimes in the evenings they took long walks about the city, discussing their ideas about the meaning of life, not going home till the early hours of morning. His friend had a great reverence for religion, whereas Erwin at this time had a vehement dislike for it, and thought that Fränzel's "religious teachers had high handedly torn him from his true way of life," presumably by inhibiting his approach to biological problems.

Erwin and Fränzel read together and discussed in detail the book of Richard Semon, *The Mneme as Conservative Principle*, first published in 1904. This was the only time that Erwin ever read a book in this way with another person. The book had a major influence on the development of his philosophical ideas, but it is difficult to discern all the values he must have found in it. Perhaps his friend Fränzel was keen on the book and communicated his enthusiasm to Erwin. Years later, when he was reading philosophy on the Italian front, he mentions Semon again as one of the important influences on his thinking. His book *What is Life?*, which had such a seminal effect on molecular biology, must have had its origin in those midnight discussions with Fränzel about the theory of living organisms.

Richard Semon took his doctorate in Jena in 1883, as the favorite pupil of Ernst Haeckel, from whom he learned the fascination of the wide-ranging theory that tries to explain everything. The basic philosophy of *The Mneme*, however, was derived from Lamarck. Erwin sold almost all his father's books after he died, an action motivated by financial need, but one that he came to regret deeply. Among the few that he kept was *Philosophie Zoologique* by Jean Baptiste Lamarck, originally published in 1809. Lamarck is famous as the founder of the doctrine of inheritance of acquired characteristics, although he was never dogmatic about this. Followers of Lamarck were divided into mechano-lamarckists and psycho-lamarckists. A modern example of the former was Trofim Lysenko. The *Just So Stories* of Kipling include several lamarckian fables, such as how the elephant got its trunk. The question is to what extent changes in the phenotype produced by environmental conditions can affect the genotype, the hereditary information transmitted to the offspring. The consensus today is that such effects have at most a minor role in the hereditary process. In most of *The Mneme*, Semon appears to accept lamarckism, but in a final chapter he praises the Darwinian selective mechanism of evolution.

Psycho-lamarckism presents a more mysterious theory, which may be why it was so interesting to Erwin and Fränzel as late adolescents. When an organism reacts with its environment, a memory trace is produced in its mind, and this memory can be transmitted to its descendants, either through an effect on the germ cells or directly through some form of psychic inheritance. Instinctual behavior is said to be an example of this inherited memory. A similar idea had been adumbrated earlier by Ewald Herring, professor of physiology at Prague and a colleague of Ernst Mach. It was also given some currency in England by Samuel Butler, the author of *Erewhon*, in his book *Unconscious Memory*. An eloquent modern exponent of psycho-lamarckism was the Swiss psychiatrist Carl Jung.

Semon's basic idea was that all biological phenomena can be ascribed to the existence in living cells of a sort of memory record called the *mneme*. Thus heredity, differentiation, regeneration, development, instinctive and learned behavior, conscious recall of past events, motor skills – are all due to the mneme. Both the inherited and the acquired characteristics of an organism are controlled by *engrams* located within its cells. Thus the regeneration of the limb of a newt is due to the engrams for limb formation in its cells, and the memory of Capri evoked in a traveler by the smell of hot olive oil is due to the corresponding engrams in his brain cells. The sum of all the engrams that an organism has inherited or acquired is its *mneme*. The production of an engram is caused by a stimulus (*Reiz*), which can be envisaged as an alteration of the energy state of the organism.

Why did Erwin find this book so fascinating? One can suggest two reasons. Firstly, he had no formal training in biology; it was a *terra incognita* in which bizarre concepts might be encountered without dismay. Secondly, and perhaps more important, all German-speaking scientists have been imbued with the spirit of Goethe, the greatest *Naturphilosoph* of all time. No matter how cold and abstract their university training, they absorbed in their youth Goethe's feeling for the unity of Nature:

Faithful observers of Nature, even if they think very differently in other things, nevertheless all agree that everything that appears, everything we meet as a phenomenon, must mean either an original division that is capable of reunion or an original unity that can be split and in this manner display itself. To sever the united, to unite the severed, that is the life of Nature, that is the eternal drawing together and relaxing . . .

Thus as Erwin and Fränzel walked about the streets of Vienna, with the gentle gas lights sliding their shadows along the pavements, we can imagine that they were inspired by more than Semon's far-fetched analogies, one boy very religious, the other very agnostic, they were seeking together an answer to the riddle of existence.

THE STUDY OF PHYSICS

The main focus of Erwin's interest at the university was the course of Hasenöhrl on theoretical physics. This extended over eight semesters with five hours a week of lectures. Hasenöhrl was youthful, full of energy, and a brilliant lecturer. He lectured without notes, but he had not memorized anything, he simply relied on the strong logic of the science and developed it as he went along. Nor did he lose sight of his aim in a thicket of mathematical details, a fault of some theoreticians. His enthusiasm for the beauty of his subject inspired his students; even if Erwin were not already dedicated to theoretical physics, these lectures would have converted him. Erwin always said that he had difficulty learning from books, and to have his subject presented with deep understanding in such wonderful lectures was an intellectual joy of the highest order. "No other person has had a stronger influence on me than Fritz Hasenöhrl, except perhaps my father." There was a certain air of chivalry about Hasenöhrl, yet his friendliness overcame any barriers of seniority between him and his students. He often had groups of them to his house, where his beautiful wife Ella presided and his small son and daughter added to the happy atmosphere. He was a strong mountaineer and expert in skiing and other winter sports. He organized expeditions with the students and took an interest in student affairs, and as Hans Thirring reported "wherever he went he acted as an energizer and brought good fellowship."

Hasenöhrl lectured in the old rented building on Türkenstrasse. The lectures covered the foundations of analytical mechanics, the dynamics of deformable bodies, the solution of partial differential equations and eigenvalue problems, Maxwell's equations and electromagnetic theory, optics, thermodynamics, and statistical mechanics. The excellence of the lectures made the students forget the dilapidated lecture room in which they were given. There were no proper benches and they had to sit on chairs and hold their notebooks on their knees. "The floor was an ancient inlaid one, through which gaping crevasses ran, in which even today untold amounts of mercury might remain. Each step made the entire room shake . . . and even the outer walls trembled when a strong wind blew outside or a truck passed by in the street."

The lectures on mathematics (function theory, differential equations, and mathematical statistics), given by Professor Wilhelm Wirtinger, were uninspired, but the material was essential for any future work in theoretical physics. Schrödinger spent many hours of many days working on mathematics at the university, and his mathematical ability achieved the first rank among contemporary theoretical physicists, perhaps surpassed only by that of Arnold Sommerfeld. Nevertheless there were some branches of mathematics, especially in algebra and group theory, that he neglected. Fortunately what he studied turned out to be what he would need for his great work. The "bible" of mathematical physics at that time was *The Partial Differential Equations of Mathematical Physics* by Georg Riemann and Ernst Weber, which he mastered in detail as a student.

The course in Meteorology that Schrödinger took with Julius Hann (1839–1921) was to prove an unexpected blessing during the First World War, when he was transferred in 1917 from the artillery on the Italian front to the Military Meteorological Service in Vienna. The chemistry course, which included practical exercises in qualitative analysis, was well taught, but Schrödinger never showed any interest in this subject.

Several members of the staff of Professor Exner were among Erwin's teachers at the Physics Institute. Egon Schweidler and

Hans Benndorf were Exner students who worked on atmospheric electricity. Erwin's closest friend among the staff was Karl Wilhelm Friedrich Kohlrausch, always called Fritz, who came from a family of notable German scientists. He was just completing his first degree while Erwin was a second-year student. Kohlrausch's thesis was an experimental study of fluctuations in the rate of disintegration of a radioactive element, a phenomenon discovered by Schweidler the year before. In 1910 he investigated the mobility of radium-A atoms in air, a problem that Schrödinger was to take up later.

On the sixtieth birthday of Hans Benndorf, Stefan Meyer and Fritz Kohlrausch had the happy idea of commemorating the Exner circle with a beautiful silver loving cup, which would be passed from one member to the next as each had his signature inscribed on it on his sixtieth birthday. It bore the invocation: "Travel O Cup and carry much gladness from one to another. Whoever drinks from you, let him think of the sexagenarian who came before and the one who comes after him." Benndorf was the oldest and Schrödinger the youngest, and after his death the cup was given into the keeping of the Academy of Sciences.

The world of Vienna outside the University continued to provide its usual distractions, but the turn-of-the-century flowering of artistic and musical talent was beginning to fade. In December 1905, Franz Lehar's *The Merry Widow* opened at the *Theater an der Wien*, and its lilting waltz became the theme song of pre-war Vienna. In 1906, Enrico Caruso came to the Opera for the first time; Musil published his novel *Young Torless*, a dire premonition of the sadism and perversion of the Nazi era; and Buffalo Bill Cody brought his wild west show to the Rotunda of the Prater. In 1907, Gustav Mahler was dismissed as director of the Opera, and a daring Kabarett, Fledermaus, was opened at Kärntnerstrasse 33, a few doors from Grandfather Bauer's house. During the next year the Wiener Werkstätte published Oskar Kokoschka's *The Dreaming Boys*; and Erwin Schrödinger became twenty-one years old and experienced his first mature love affair.

Little is known about this early love, except her name, Ella

Kolbe, and the fact that Erwin fell in love with her that year and a short but intense affair followed. He was still living in the family home, but another physics student, Jakob Salpeter, with whom he shared a laboratory, had a small apartment near the university where Erwin sometimes stayed and where the lovers might meet. He also continued to visit the Rella family occasionally, but he made it rather clear to Lotte that he was not interested in marriage at this time, despite any hopes her mother may have had for their engagement.

DOCTORAL DISSERTATION

In the Austrian universities, the first degree in physics was the Doctor of Philosophy, Ph.D. In addition to completing the necessary courses, the candidate was required to present an original dissertation to the faculty. After the dissertation was accepted, there were final exams called *Rigorosa*, one of which was about some philosophical topic, as a concession to the name of the degree. The degree was not equivalent to a modern Ph.D. in physics, which typically requires at least three years of advanced study and research beyond the first degree. The Austrian doctorate was at about the level of a Master's degree at an American university or an Honour's degree in Britain. Schrödinger was awarded the Ph.D. degree on May 20, 1910.

His dissertation was "On the conduction of electricity on the surface of insulators in moist air." The problem was motivated by the importance of electrical insulation in instruments for the measurement of radioactivity and ionization. He thanked Professor Franz Exner in whose department the work was done and acknowledged the support of (Associate) Professor Schweidler.

Schrödinger's first research displayed considerable experimental ability but, surprisingly, it lacked any theoretical content. It was a routine set of electrical measurements designed to show the effects of moist air on the conductivity of several solid insulators; ebonite, amber, glass, sulfur, and paraffin wax. He assumed that the effects were restricted to the surface but neglected to make the obvious control experiment to prove this, namely, variation of the area at constant cross section. His practical experience with these insulating solids (called *dielectrics*) did have one important consequence: it was the origin of the masterly survey of dielectricity that he completed four years later. Nevertheless, the dissertation, even considering that it was a first effort, was not a remarkable performance, and one cannot discern any sparks from the "fire spirit" portrayed by Hans Thirring. The work was of sufficient interest, however, to be presented at a meeting of the Vienna Academy of Sciences on June 30, 1910. This was the first occasion at which Schrödinger delivered a report on his own research work to fellow scientists. It was published a few weeks later in the *Proceedings* of the Academy.

MILITARY TRAINING

Austria-Hungary required universal military service: three years service from all able-bodied young men. Since the large number of reserve formations needed more officers than could be provided by the military academies, the Empire adopted the Prussian system of "one year volunteers" (*Einjahrige Freiwillige*). Men of adequate education and social standing were allowed to volunteer for one year's training as officer candidates, and after passing an examination, were commissioned in the reserves. The divisions of the army differed in social prestige, the infantry ranking lowest, the cavalry highest, and the artillery in between, the fortress artillery being more snobbish than the field artillery.

On October 1, 1908, Erwin enrolled for military service, and on June 6, 1910, soon after he graduated at the university, he was accepted as an *Einjahrig Freiwillig* in the fortress artillery. On October 1, 1910, he presented himself for active service. The personal description entered on the first page of his record book noted that he had blond hair and blue eyes and blond eyebrows, but somebody later crossed out "blue" and wrote "green?" His height was given as 167.5 cm (5 ft. 6 in.).

Volunteers were responsible for their own living expenses.

The first two months of training were spent in barracks, but after that they were free to find board and lodging in accord with their means in the towns in which they were stationed. Life in the garrison towns could be boring, but there were also happy evenings of wine, women, and song for those so inclined. Schrödinger enjoyed his assignment in the beautiful city of Krakow. Although the dress uniforms of the cadets were indeed elegant, many of the one-year trainees never bothered to learn enough military science to pass the final examination. For many young intellectuals, however, a year of outdoor exercise and good fellowship was a welcome change from a sedentary life amid dusty books.

Erwin had leave over Christmas and he went with Hans Thirring for a ski course at Mariazell, in the mountains south-west of Vienna. Hans broke his foot while ski-jumping. He was in considerable pain, and Erwin stayed with him all night, moving the foot to a more comfortable position whenever the pain became intolerable. Erwin had planned a long ski tour for the next day, but he did not consider his own need for sleep, only the care of his friend.

Schrödinger's military service had an important influence on his career. When he graduated there was an assistantship in theoretical physics available with Hasenöhrl, and he would naturally have been appointed, being first in his class. Since, however, he was not available, the appointment went to Thirring, who was excused from military service as a result of his ski accident. Erwin completed his military training, passed the examination, and received his commission as *Fahnrich* in the reserves, a rank just below that of lieutenant. On New Year's Day, 1911, he returned to civilian life. There is no evidence that he did much physics in the army, but his health was undoubtedly improved.

UNIVERSITY ASSISTANTSHIP

Schrödinger was appointed to an assistantship in experimental physics in the department of Franz Exner and under the supervision of Fritz Kohlrausch. He was put in charge of the large practical class in first-year physics, a sometimes tedious task, but one upon which he embarked with enthusiasm. As he recalled later:

I learned two things during these years: First, that I myself was not suited to be an experimentalist. Second, that the land in which I lived, and the people with whom I lived there, were no more suited than I to achieve experimental progress along major lines. This was mostly a consequence, among other things, of the tendency of the golden Vienna heart to place amiable nincompoops in key positions (often only on the basis of seniority), where they blocked progress, while active personalities were needed there, who would have had to be brought in from outside. Thus atmospheric electricity and radio-activity, which really had their beginnings in Vienna, were taken out of our hands, and anyone who felt inspired to work seriously in these fields had to go abroad, as for example Lise Meitner from Vienna to Berlin.

Schrödinger was happy to have free access to the excellent collection of optical instruments, so that he was able at any time to make observations of spectra and interferometric measurements. He also had facilities for mixing light of different colors and recording the resultant hues, intensities, and saturations. In this way he discovered that his own color vision was that of a deuteroanomalous trichromat; the cone cells of his retina included those for the three colors red, green, and blue, but the relative intensity of his red perception was enhanced. This anomaly in color vision occurs in about two percent of all those tested.

From his laboratory experience, Schrödinger was able to say "I belong to those theoreticians who know by direct observation what it means to make a measurement. Methinks it were better if there were more of them." When the "measurement problem" became the subject of a great controversy in quantum mechanics, he was able to hark back to his youthful days in the physics laboratory, where he had learned that physics is not based on mathematical fantasies but on a solid ground of experimental observations. Thus his work as a laboratory assistant helped to determine the philosophical framework that he was willing to accept for a physical theory.

HABILITATION: FIRST THEORETICAL PAPERS

As soon as his teaching duties were well organized, Schrödinger began to consider the original research that would allow him to ascend the next step of the academic ladder. This was the *Habilitation* or admission to the *Venia Legendi*, which allows one to serve as *Privat Dozent* at the University. The *Dozents* were paid only nominal salaries, but could offer courses for which they received fees. *Habilitation* required the demonstration of ability to do original research as evidenced by publication of papers. These papers were considered by a committee of professors, along with the overall record of the candidate and his plans for courses. He was also expected to give a special public lecture attended by members of his committee.

Even for those who could fulfill these requirements, there were few good academic opportunities in Austrian physics. Hans Thirring recalled one discussion with Erwin. "In contrast with other young titans, he maintained an astonishing inner modesty. Once, in 1911, we were talking about the not so rosy career prospects for physicists. I mentioned that one might perhaps find a position in Germany if all the local ones were filled. Schrödinger stood still for a moment, shook his head, and said 'Yes but then one must already have accomplished something quite special'." Those who knew Schrödinger divide into two distinct classes: those who considered him to be a person of amazing modesty, and those who thought he was one of the most conceited men they had ever met, with the majority being of the latter opinion.

For his *Habilitation* papers, Schrödinger chose theoretical problems related to the interests of Franz Exner. His first theoretical paper was presented before the Vienna Academy on June 20, 1912, "On the Kinetic Theory of Magnetism." He based his work on an important 1905 paper by the Dutch theoretician Lorentz, which pictured a metal as an array of positively charged ions permeated by freely mobile electrons called an *electron gas*. Schrödinger's idea was to calculate the magnetism due to the electron gas. At the very beginning, he fell into an error fatal to the final result, by assuming that the electrons would follow the Maxwell law for their velocity distribution and thus have the same average kinetic energy as a gas molecule at the same temperature. Although he began with a false premise, Schrödinger developed his mathematical theory clearly and elegantly. Not surprisingly, the equation he derived did not agree with the experimental data. When he presented this work to his *Habilitation* committee, Fritz Hasenöhrl gave it a somewhat equivocal commendation.

For his formal *Habilitationschrift* Schrödinger chose his paper "Studies on the Kinetics of Dielectrics, the Melting Point, Pyro- and Piezo-Electricity" which was presented to the Academy on October 17, 1912. He was attracted to the fundamental problem of why and how a solid melts to a liquid. It is not easy to devise a satisfactory atomic theory for the melting transition from an ordered anisotropic crystalline state to a disordered isotropic liquid state, and even today available theories leave much to be desired. As he wrote:

How do real solid substances get into this well ordered structure? Why does it suddenly disappear at a definite temperature? I believe I can give a sort of atomistic answer to these questions. It is still in very unfinished form with many and great gaps. Above all it lacks the most important and the most difficult parts: the equations of state . . . But that is unfortunately future music [*Aber das ist leider Zukunftsmusik*].

Again Schrödinger displayed superb mathematical facility combined with a lack of insight into the physical realities of the problem. Although he was working closely with experimental physicists, neither he nor they had yet mastered the basic approach of modern science, the design of experiments with the specific purpose of testing a theoretical model. To them, experimental physics consisted in making careful measurements, a collection of data which would later be subjected to theoretical analysis. Theoretical physics consisted in making a model, usually based on electromagnetic concepts, and then using mathematics to derive relations between certain properties. The idea of a close co-ordination in which theory guides experiment and experiment tests theory was not generally understood at the time.

In the brief report of the *Habilitation* committee on this work, Hasenöhrl did not deal with it critically. There is evidence here of the parochial character of Viennese physics. Schrödinger's early scientific development was inhibited by the absence of a group of first-class theoreticians in Vienna, against whom he could sharpen his skills by daily argument and mutual criticism. The committee was not unanimous in recommending acceptance of Schrödinger's application for *Habilitation*. Professor Wegscheider wrote that "My opinion is inclined against this premature *Habilitation* on the grounds of the youth of the candidate and the small amount of research work completed." But Schrödinger had the support of Hasenöhrl, and at the faculty meeting of June 14, his application was approved by 38 votes to 2, with 5 abstentions. After further formalities, the Dean was able to send the nomination to the Imperial Ministry for Culture and Instruction, and on January 9, 1914, the Minister sent his approval. Thus Erwin Schrödinger became a *Privat Dozent* for Physics at his Alma Mater. He was standing on the first rung of the academic ladder.

FELICIE

Just as he was taking his first forward step in his academic career, Erwin was strongly tempted to abandon it altogether for the sake of a young woman. Among the friends of Rudolf and Georgine Schrödinger were Karl and Johanna Krauss. They were a fairly wealthy family with connections to the minor nobility and a long tradition of strict observance of the Catholic faith. Karl was trained as a lawyer and worked in the Imperial Archives. Their daughter Felicie was born in 1896, and was therefore about eight years younger than Erwin. As a boy, he used to be told at children's gatherings, "now look after little Felicie," a task that he undertook with considerable reluctance. When her father died in 1911, Felicie was not yet fifteen years old. She was growing up to be a beautiful teenager, and Erwin was now delighted to be with her as much as possible. He was always captivated by girls who were just at the onset of womanhood. Felicie's mother, however, wanted to discourage any closer relationship with the Schrödingers and viewed Erwin as a quite impossible match for her daughter. With the support of her relations, she issued an edict that Erwin and Felicie would be allowed to meet only once a month. As might be expected, such separation made their hearts grow fonder, and they were soon deeply in love. They wanted to get married and considered themselves to be informally engaged.

Erwin's prospects as a physicist were dismal from a financial point of view. At best he could hope to spend several years as a *Privat Dozent* with no assured income or to have an assistantship with a miserable salary. He was desperately in love with Felicie and went to his father to ask if he could give up the university and take a leading role in the family linoleum business. He wanted to get married and saw no likelihood of being able as a physicist to support a wife. With great wisdom, Rudolf said no, he would not allow such a thing. He had sacrificed the scientific work he loved to go into business and had regretted it all his life; he did not intend to see his son make the same kind of mistake. Erwin must stay with his university work in physics. Besides the family factory was not prospering and would not really provide for Felicie to the extent her family would think appropriate.

Erwin was now twenty-five and Felicie seventeen. Baronin Krauss became alarmed at their developing relationship. Her most serious objection to Erwin as a son-in-law was probably not that he was poor. The social standing of his parents was a greater obstacle, for the Krauss family as members of the minor nobility could place a "von" before their surname, which was important in imperial Vienna. Even worse was the fact that Erwin was nominally a Protestant, but actually a *Freidenker*, a free-thinker who probably did not even believe in God. A noble Protestant might have been acceptable but a poor atheist dabbling in arcane mathematics was unthinkable as a son-in-law.

It would have been a most unsuitable match for many reasons, and Felicie was persuaded (or forced by her strong-

7. Felicie Krauss (c. 1913)

willed mother) to break the informal engagement. This happened about the middle of 1913. In the context of Austrian society at that time, it would have been virtually impossible for Felicie to withstand the pressure of her family and marry Erwin. Yet the result was an incalculable loss for Erwin. His earlier loves had not been superficial, but he did not think of them as permanent commitments, whereas he was willing to dedicate his emotional life to Felicie. Probably at this time he began to form his negative view of bourgeois marriage. Since he was prevented by social pressures from a dedication of his spiritual, romantic, and sexual longings to the one person of his choice, he would henceforth look with disdain upon the institution of marriage and attempt to construct his emotional life outside its rigid framework.

After the breaking of his engagement to Felicie, many years would pass before Erwin again fell in love with a woman who belonged to a social class equal or superior to his own. In view of the shallowness of most Viennese religious convictions, he may well have ascribed his rejection by the Krauss family to social rather than ecclesiastical factors, and perhaps inwardly he resolved never again to suffer such a blow to his pride. This interpretation would help to explain the kind of woman he decided to marry, and his subsequent treatment of his wife as a sort of superior domestic servant. He would never have treated Felicie in such a way.

It must be admitted, however, that the harsh policy of Felicie's mother probably prevented future unhappiness for both young lovers, but one cannot be sure – if Felicie had married Erwin and given him a son, he might have been satisfied to recreate the pattern of his own family circle. One side of his personality was seeking a tranquil life of teaching and philosophical meditation, yet he may have needed the excitement of tempestuous sexual adventures to inspire the ardent creativity that produced his great theoretical breakthroughs.

In 1917, Felicie married a young lieutenant in the Austrian army, Ferdinand Bianchi. He was a baron like her father and came from an old imperial military family, which at that time was quite wealthy, although later they lost their money. He was thus eminently agreeable to Felicie's mother on religious, social, and financial grounds. Felicie never lost her interest in Erwin's career and later became a good friend of his wife. She sent Erwin a long poem for his seventieth birthday, including the stanza,

> Mir scheint, es ging bei ersten Hahnenschrei
> Grad unsre Kindheit eben erst vorbei,
> Mir scheint, als war es nur ein Augenblick
> Seit unsrer Jugend Ubermut und Glück.

It seems to me, by the first cock's crow, / Our childhood time is only just gone by, / It seems to me, as if but a moment has passed / Since the playfulness and joy of our youth.

ATMOSPHERIC ELECTRICITY AND RADIOACTIVITY

One of the major research subjects in the Physics Institute II was atmospheric electricity and Schrödinger would be expected to have an interest in this field. Egon Schweidler had established a number of observational stations to study the electrical conductivity of the atmosphere, at Attersee, Ossiachersee, and the principal one at Seeham on the Mattsee near Salzburg. Perhaps there was a scientific reason for favoring such lakeshore resort towns, but they were also exceptionally pleasant places to work during the summer. Schrödinger's first paper on atmospheric electricity was published as Number 48 of a series of seventy-eight from the Physics Institute.

For many years physicists had been puzzled by the origin of the background radiation that caused a slow discharge of electroscopes. Two sources of radioactivity were found, one in the surface layers of the earth due to radium, thorium, and their decay products, and the other in the atmosphere due mainly to radium emanation (radium-A, now called radon) and its decay products. Schrödinger's Vienna colleague, Viktor Hess, was on the track of a much more exciting possibility: an extraterrestrial source of high-energy radiation. Hess was an ardent balloonist and in April 1912, he began a series of ascents with precision electroscopes. On August 7, a flight reached 4,800 meters where the discharge rate was about three times that at ground level. Hess published his data in November 1912, with the conclusion: "The results of my observations are best explained by the assumption that a radiation of very great penetrating power enters our atmosphere from above." Thus the discovery of cosmic rays was recorded, although they were so named by Robert Millikan only some years later, and Hess did not receive his Nobel prize till 1936. Schrödinger derived an expression for the variation with altitude of the atmospheric radioactivity arising from sources in the soil, a background with which the observations of Hess could be compared.

In August 1910, Kohlrausch had made a series of measurements of the radium-A content of the atmosphere at Seeham. Professor Exner suggested to Erwin that it would be a good idea to repeat the Kohlrausch measurements at the same place in the summer of 1913. Schrödinger set up his equipment near the window of his hotel room, about 200 meters from the location used by Kohlrausch. Between July 24 and September 5, he made 229 determinations, certainly earning his spurs as an experimental physicist in this painstaking work. His activity values were only about one-fifth those found by Kohlrausch. He states

I have often explored this remarkable result in detail with my friend Kohlrausch . . . There remains no other explanation than that the yield of emanation from the pores of the soil was strongly reduced as a consequence of the repeated and thorough soaking of the soil in the rainy summer of 1913, which seldom left us even one week long without rain.

Schrödinger, though primarily a theoretician, was able and willing to contribute to the communal effort of the Physics Institute on atmospheric electricity. The subject was more important than its devotees probably realized, since we now know from the work of Hermann Muller and others that an important factor in determining the mutation rate of genetic information (DNA) is the high-energy radiation from both terrestrial and extraterrestrial sources.

Erwin loved the outdoors and the Austrian countryside. As a

compensation for slaving away at his electroscope, he was able to enjoy hiking in the surrounding hills and swimming in the cool lake. The Kohlrausches came with their children to spend a holiday and they had many happy excursions with Erwin. More importantly, they introduced him to a pretty and jolly teenage girl, Annemarie Bertel from Salzburg, who was helping take care of the children. Dressed in the local costume, she looked like a peasant girl, but her father was court photographer and a man of some substance in Salzburg. Anny's birthday was New Year's Eve (*Sylvesterabend*) 1896. She was enormously impressed by the young scientist and thought "he was very good looking." At that time her meeting with Erwin did not blossom into a serious love affair, since Anny was a country girl, "practically a child" in dirindl and pigtails, while Erwin was a serious academic, with all the sophistication of the capital city. Besides, he had not yet recovered from the loss of his true love Felicie. Yet it was more than a holiday romance, as the future would tell.

On October 1, 1919, Erwin gave Anny a copy of the 1913 paper with an addendum written in a parody of his scientific style:

As I now first discover, there must have been at that time something else of a different nature in the air at Seeham, besides Radium A, B, and C, of which, however, my electrometer did not detect a trace. The credit for the discovery is due exclusively to Fräulein Bertel (Salzburg) who drew the attention of the author to it . . . A joint publication of the above-mentioned discoverer and the author will follow in another place in the near future.

When his last measurements were made, Erwin quickly packed his equipment, and rushed back to Vienna to take part in preparations for an important international scientific congress, the eighty-fifth Meeting of German Scientists and Physicians.

THE 1913 CONGRESS OF VIENNA

Before the previous Vienna meeting of this important society in 1894, Franz Exner had left the city because he was so ashamed of the miserable quarters of the physics institutes. Now he was proud to welcome his confrères to his beautiful new building. Twenty years had elapsed since he had accepted the Vienna chair with the understanding that a new building would soon be provided to replace the rickety quarters on Türkenstrasse, but it was not completed until the spring of 1913. The site faces Boltzmanngasse, next to the Institute for Radium Research and about a kilometer from the main university building on the Ring.

More than 7,000 participants attended the meeting, which included lectures on a wide range of scientific and medical subjects. The imperial city had never looked more splendid than in these last autumn days of its glory. The festivities included a reception at the imperial court and a banquet given by the city of Vienna in the *Rathaus*. In addition, there were social activities for the physicists and their wives, including a party with music and dancing in the halls of the new Physics Institute.

The most important lecture for the physicists was scheduled for Tuesday morning, by Albert Einstein on "The Present Status of the Problem of Gravitation." He was then a professor at the Technical University (ETH) in Zurich but was soon to move to Berlin. For his work on special relativity and quantum theory, he was recognized as the greatest theoretical physicist in the world, although he had not yet received a Nobel prize. The lecture room was filled to capacity when he began to speak at 9:00 a.m.

Einstein did not overwhelm his listeners with awesome mathematics. He began with the most simple and basic ideas and developed an analogy between gravitation and electrical theory. The latter began with electric charges and the inverse square force between them given by Coulomb's law. Later electric currents were discovered and it was found that they could be generated by moving magnets. Then, with the work of James Clerk Maxwell and Heinrich Hertz, electromagnetic waves were discovered and optical and electromagnetic phenomena became parts of a unified field theory. Our understanding of gravity, he said, is now only at the stage of Newton's law of inverse square attraction between two masses. It is necessary to modify the instantaneous action at a distance of Newton's theory in order to exclude any physical influences that move faster than the speed of light. Einstein outlined the work that he and Marcel Grossmann had been doing in Zurich. Their theory, unlike others, required a bending of light rays in a strong gravitational field, and this prediction could be checked by observing stars appearing close to the sun during a solar eclipse. "Let us hope that the eclipse of the year 1914 will allow us to obtain this important decision."

Einstein's theory was based on the identity of gravitational and inertial mass, and the doctrine of Ernst Mach that inertial mass is due to the effect of other masses in the universe. While in Vienna, Einstein visited Mach, who seldom left his apartment. Phillip Frank recalls that on entering Mach's room, "one saw a man with a gray, unkempt beard and a partly good-natured, partly cunning expression on his face, who looked like a Slavic peasant." Einstein asked Mach to suppose that he could predict by atomic theory an observable property that could not be predicted by continuum theory, would he then accept that it was "economical" to assume the existence of atoms? Mach admitted that the atomic hypothesis would then be economical, and Einstein went away quite pleased.

Schrödinger's interest in gravitation was kindled by this brilliant lecture, but he must have filed it away at the back of his mind until Einstein's great papers on general relativity were published in 1915, and he had time to study them while in charge of an artillery battery on the Italian front. Also the idea of a unified field theory, which would include both electromagnetism and gravity, must have occurred to some of the young scientists listening to Einstein's outline of the analogies between the two fields. In their mature years, both Schrödinger and Einstein would devote their lives to the quest for this unification.

Schrödinger did not lecture at the meeting; he had not yet become a *Privat Dozent* and besides he had nothing special to report. The brilliant array of German physicists impressed him strongly. The work of Max Laue on X-rays and crystal structure made a particular impact, for he had begun some work on the effect of temperature on X-ray diffraction. When he received his appointment as *Dozent* a few months later, his first course, offered for the summer term of 1914, was on "Interference Phenomena of X-Rays."

X-RAY AND DIELECTRIC STUDIES

Earlier in 1913, Pieter Debye had published two papers on the effect of the thermal vibrations of the atoms in a crystal on its X-ray diffraction pattern. Although he must have known that Debye was still working on the problem, Schrödinger took it up also. As had been typical of his theories so far, his mathematical treatment was highly polished, but his physical model was less than adequate. The importance of this work for Schrödinger's future research was that it directed his attention to problems of atomic vibrations and scattering of radiation, in which the quantum theory must have an important role. These problems led him away from the special interests of the Vienna School and brought him closer to the mainstream of European atomic physics.

THE FIRST OUTSTANDING PAPER

Early in March 1914, Schrödinger sent to the *Annalen der Physik*, the most important German physics journal, a purely theoretical paper, "On the Dynamics of Elastically Coupled Point Systems," which is the most significant of his prewar publications. This paper is especially interesting in that it harks back to one of the persistent themes in the scientific life of Erwin's intellectual grandfather, Ludwig Boltzmann, i.e., the absolute necessity for atomistic models in physical theory. Much of theoretical physics is expressed in the form of differential equations, which describe the trajectories in time and space of quantities such as temperature, electric-field strength, concentration. The equations are completely general, and to apply them to specific systems, they must be integrated subject to certain initial conditions (at time zero) or boundary conditions (values of the quantities of interest at the boundaries of the system). Differential equations try to represent the macroscopic behavior of a system without reference to its underlying atomic structure. For example, one might consider an idealized string with a certain elasticity and mass per unit length, and calculate how it vibrates when subjected to some initial disturbance. Advocates of strict phenomenalism, such as Ernst Mach, contended that such differential equations simply represent in mathematical form the world of direct experience. In opposition to this view, Boltzmann argued that a continuum in physics is not something truly continuous, but rather a finite number of elements allowed to increase in number until further increase no longer has any effect, so that "you cannot disengage yourself from atomistics by the consideration of differential equations."

As an ardent student of Boltzmann's work, Schrödinger would have been well aware of the above remarks of the master, when he began his own extension of the subject:

It is so to speak part of the creed of the atomist that all the partial differential equations of mathematical physics . . . are incorrect in a strictly mathematical sense. For the mathematical symbol of the differential quotient describes the transition in the limit to arbitrarily small spatial variations, while we are convinced that in forming such "physical" differential quotients we must stop at "physically infinitely small" regions, i.e., at those that still always contain very many molecules; if we were to push the limiting process further, the quotients concerned, which up to then really were proceeding nearer and nearer toward a definite limit . . . would again begin to diverge strongly.

Thus the limit used for the differential equations is only a "pseudo-limit" and is not the same as the limit conceived by pure mathematics.

Atomistics has the task of finding and stating those conditions under which the differential equation based on a continuum actually leads to an incorrect result because of the truly atomistic structure of matter. Schrödinger investigated a model that had recently been applied by Max Born and Theodore Karman to the specific heat of solids: an infinite

8. *Privat Dozent* (c. 1914)

linear array of mass points separated at equilibrium by the same distance, each mass being subject to an elastic restoring force when displaced from its equilibrium position. He derived an equation for the resulting motions of the mass points.

"As the simplest example of the sometimes astonishing results which this formula yields with great facility," he considered the following case. At $t = 0$ all the mass points are assumed to be at rest except one, which is given an arbitrary displacement. All the mass points then begin to move, but their motion is nothing like that of a continuous string as derived from the partial differential equation. He then asks the question, "What must be the character of the system of initial values in order that the motion of our discrete model should really display a certain similarity to the motion of a continuous elastic medium?" Schrödinger shows that one can apply the differential equations to such atomic models if one does not consider the variables themselves, but averages of the variables over regions that contain many atoms. For example, slow acoustic waves can be considered without reference to effects due to rapid thermal motions of the individual atoms.

This paper is undoubtedly the most interesting of all those written by Schrödinger before he was called into military service in 1914. It carries forward one of the basic problems of "grandfather Boltzmann" by its penetrating analysis of a system based on an atomic model, and it forms a bridge to his later revolutionary applications of the differential equations of wave motion in his wave mechanics. In this paper also, for the first time, we hear the authentic Schrödinger style, with its urbane confidence and its ability to relate the question in hand to deeper philosophic concerns of mathematical physics.

CHAPTER 3

Schrödinger at war

Schrödinger began his career in physics during the last peaceful years of the Danube monarchy. Old Franz Josef still retained the respect of the lower and upper classes among his subjects, but his control over the bourgeois politicians and their nationalistic tendencies began to falter. The Hungarian leaders consistently placed the privileges of the great Magyar landowners above all else, and extorted special concessions as the price of loyalty to the throne. The annexation of Bosnia-Herzegovina in 1908 created a large population of disaffected Slavs and Muslims in the southern borderlands. The K & K structure was becoming more unstable, and as for a radioactive atom, the question was not if but when it would disintegrate.

BEFORE THE WAR

The chief of the general staff, Franz Conrad, was a persistent warmonger, often advocating a "preventive war," one year against Serbia, the next against Italy. At the end of 1911, he was dismissed, not so much for meddling in foreign policy as because of his affair with a married woman, Gina Reininghaus, offended the religious sensibilities of the old emperor. After a year in disfavor, he was called back, but soon thereafter Franz Ferdinand, heir to the throne, was appointed General Inspector of all the Armed Forces.

Schrödinger took no serious interest in the political situation of the empire. His physics, his intensive reading of philosophy, the theater, and his love affairs left no time for politics. His friend Hans Thirring, however, was becoming a convinced

63

socialist and pacifist, and was fortunately exempt from military service owing to a skiing accident.

As was customary at that time for unmarried sons and daughters, Erwin lived at home with his parents. After the break with Felicie, he fell in love again, with Irene Drexler. Once again it was a romantic love that was never to achieve fulfillment. The fact that during his middle twenties Erwin fell in love with three young women but slept with none of them suggests that Vienna maidens from good middle-class families did not then indulge in love affairs, and that Erwin's other amorous pastimes were enjoyed without romantic complications.

THE DRUMS OF AUGUST

On June 28, 1914, Franz Ferdinand, resplendent in a white tunic and a general's head-dress of green cock feathers, and his beloved but morganatic wife, formerly Sophie Chotek, lady-in-waiting, in a picture hat and a high-necked white summer dress, were riding in a fine open car down the main street of Sarajevo in Bosnia. A few hours previously they had narrowly escaped assassination from a bomb. Suddenly, a young man, Gavrilo Princip, stepped up to the side of the car and fired two shots, one through the neck of the Archduke and the other into the abdomen of his wife. She died almost immediately and he died a few minutes after the cars reached the Residency. Princip was one of a group of conspirators who had been provided with weapons by the Black Hand, a powerful secret society with headquarters in Serbia.

The news of the assassination was received calmly in Vienna, for Franz Ferdinand had been generally unpopular. The old emperor was rather relieved at the removal of a somewhat untrustworthy heir, for now his great nephew Charles, a pious but predictable person, was heir apparent. The court was mainly concerned to humiliate Sophie in death as it had enjoyed doing in life. Conrad, however, knew that war was inevitable, and he wrote to Gina on the evening of the assassination: "It will be a hopeless struggle, but nevertheless it must be, because such an ancient monarchy and such an ancient army cannot perish ingloriously."

On July 31, 1914, Rudolf brought Erwin's mobilization orders to his office in the new Physics Institute. They went to buy two pistols, one very small and one very large. "Luckily," recalled Erwin, "I have never fired them at man or beast." Before reporting for duty, he sent Anny Bertel a book of essays by Felix Salten, which he inscribed "In grateful remembrance of the beautiful long days."

His orders were to report to Predilsattel, near the Italian border, a fortified position on a mountain pass about 1,000 meters high, commanding the Seebach valley that opens out onto the Venetian plain. This was an excellent place to be at the beginning of the war, well away from the murderous battle of Lemberg on the Russian front, where Conrad lost one-third of his effective troops in the first three weeks of fighting, 250,000 killed or wounded and 100,000 prisoners. There were observation posts on elevated positions at both sides of the fort, and Erwin spent some time in one of them.

One night . . . we saw a number of lights moving up the slopes at the head of the lake, where there were no paths, apparently coming towards our position. I sprang up . . . and went . . . to survey the situation. The observation was correct, but the lights were St. Elmo's fire on the points of the barbed wire entanglements only one or two meters away, the displacement onto the background being caused by parallax as a result of the movement of the observer himself. I saw this really enchanting phenomenon another time on the points of grasses on clods of earth that covered our roomy dugout . . . I don't remember ever seeing St. Elmo's fire before or since.

RESEARCH IN UNIFORM

During the first months of the war Schrödinger was able to complete some scientific work. The advantage of being a theoretician was that he could make calculations even in a dugout, although he missed the possibility of consulting the scientific journals. On October 27 he posted to the *Annalen* a short paper on "Capillary Pressure in Gas Bubbles." He had been working on this problem at the outbreak of the war; it was probably suggested by a student experiment on measurement of surface tension. Three different authors had derived three different expressions for the surface tension and Schrödinger showed that all three were wrong. He gave the correct expression and also a new formula for the bubble-pressure method. He had begun some experimental measurements and they agreed well with his equation. He concluded by saying "I intend to continue the measurements, which had to be interrupted at the end of July, so as to obtain more extensive data for testing the formula." Obviously he never thought that the war would continue four more years.

Schrödinger's next assignment was to Franzenfeste, a fortress just north of Brixen in South Tirol, which was built in 1835–38 to dominate the Brixener Klause, a gorge at the entrance of the Brenner Pass. Here he was in the mountain country that he loved, surrounded by snowy peaks. It was a perfectly peaceful sector during the winter of 1914–15. Meanwhile the German offensive toward Paris had ground to a halt and the war of attrition had begun, which would take millions of lives in the muddy trenches of the Western Front. The Central Powers had lost their chance to win the war, but the slow process of their destruction was only beginning.

A most unwelcome surprise to the forces of the Empire was a humiliating defeat by Serbia. The Austro-Hungarians, mostly reserve troops, were under the command of Oskar Potiorek, who had been in charge of security at Sarajevo. The Serbians drove into Hungary, and an alarmed Conrad planned a defensive line from Lake Baloton to Budapest. At about this time, Schrödinger was transferred to Komaron, a Hungarian garrison town between Vienna and Budapest.

While there he was able to write a paper for the *Physikalische Zeitschrift*, "On the Theory of Experiments on the Rise and Fall of Particles in Brownian Motion." This theory was related to work in progress in Vienna on the charge of the electron. One of the most important experiments in twentieth-century physics was the determination of the electronic charge by Robert Millikan at the University of Chicago in the spring of 1909. He was able to isolate an individual droplet of oil or water and measure its charge by its motion in the gravitational and electric fields between a pair of metal plates. Owing to Brownian motion, the time required for a suspended particle to traverse a fixed vertical distance in a Millikan-type experiment varies considerably from run to run, and it is necessary to average the times over a large number of observations to obtain a reliable result. The unusual statistical problem that thus rises had not been correctly solved before Schrödinger took it up in one of his most elegant early works, one that is particularly relevant to his later research because it is his first publication in statistics, a subject that lies at the heart of interpretations of quantum mechanics.

ON THE ITALIAN FRONT

On April 26, 1915, the secret "Pact of London" was signed between the Triple Entente and Italy promising her large slices of Austrian territory, control of Albania, and some colonies in Africa and Asia Minor. Italy no longer hesitated and on May 23 declared war on Austria-Hungary. This action aroused great popular feeling in the Empire. Fritz Hasenöhrl was so eager to fight the Italians that he arranged to be transferred immediately to the Italian front, where on July 20 he sustained a severe wound in an assault on Monte Piano. He was transferred to the hospital at Salzburg, but returned to the fight before he had fully recovered.

On July 26, Schrödinger's unit was ordered from Komaron to install a new 12-centimeter marine battery on the Italian Front. For the next two months, he kept a detailed diary of his activities. The marine battery was a great puzzle, for nobody seemed to know exactly where it was supposed to go, only the name of the locality, a place called Oreia Draga on the Istrian peninsula south of Trieste. He set off at 10:00 a.m. from the railway station at Wiener Neustadt. Irene Drexler was there to wish him farewell from the train platform. By noon next day he was in Marburg (Maribor) trying to arrange transport for the gun. He met the railway officer from Leibach (Ljubliana) and more or less by chance they got the necessary instructions, but the whole project was delayed since his unit did not know where they were supposed to report.

Early next morning the second Italian attack on Görz (Gorizia) began. A battle raged around the Karst plateau, where the barren limestone rocks were pulverized by artillery fire. A major Italian offensive, the first of many battles of the Isonzo, had opened on June 23. In repeated assaults the Italians failed to make significant gains through the Austrian defenses, but losses were heavy on both sides. A realistic picture of this section of the front as seen from the Italian side was given by Ernest Hemingway in *A Farewell to Arms* (1929). He did not arrive on the Italian Front till June 1918, and he was blown up by an Austrian mortar on July 8. Yet his reporting was so accurate that Mussolini banned the book in Italy.

With the gun finally installed, Schrödinger noted on August 2 that "we are firing badly. I was balled out. I took control of all the operations myself, which slowed down the action. I was balled out some more. By afternoon it was going well. It is incredible that we were not fired upon at all or only insignificantly since we were using very smoky powder. Airplanes were looking for us." On August 4, the shooting was again poor; Erwin suggested what to do but they did not believe him. "Thank God! The other battery is also shooting poorly." On August 7, troubles with the mortar continued. After six shots, the block elevation increased. They managed some good shots, but again the block elevation rose and finally the gun dug itself a hole in the earth. They came under heavy fire from at least two enemy batteries, first with shrapnel and then with heavy shells. "The men were very brave; we are generally well protected and if a shell doesn't hit the roof directly in the middle not much can happen . . . once in a while a shell hit an observer, or a dugout where four or five men were blown apart. Otherwise one should have no misfortunes."

On August 21, there is an abrupt change in the tone of the diary.

Dreamed again of L[otte] the whole night long. "The whole night" was probably five minutes, but such is the subjective feeling as one awakes. I cannot get free of her, even when I know she is not worth it, but it is as if I were still under the enchantment of that evening when for the first time I held her hand in mine, of the walks in the snow, those few happy days, which no greater favor could, or will, put in the shade. I knew long ago that the structure would tumble down, when she stretched out her hand to me from the railway platform on September 12, 1912. I had no more to hope for, might wish for no more, than the fleeting beauty of the instant. I was a child with the child, and to reawaken the vanished *Angst* of a childish eroticism was for the 25-year-old man a source of infinite charm.

For all that, I still hope, I still believe, as I curse her another thousand times, that I can call myself lucky to have escaped the foolish creature. How deeply I now consider the humiliations which I, – which she imposed upon me, uncalled for. I was foolish, I alone was foolish. I gave the reins that I could have controlled, that I disdained to control, to the child in whom I saw a goddess, which she was to me, not the child that she remained longer than many others. But even so I could have led her, must have led her.

Could have led – I? Not I, but another in my place. And yet I cannot reproach myself.

This confession shows how permanently Erwin's first love affected his soul. Although since then he had experienced a passionate affair with Ella and romance with Irene, and also the desire to make a married life with Felicie, still, when he was in the midst of the cannon fire and random death and mutilation of senseless warfare, it was his high-school sweetheart Lotte who came from his subconscious mind into his dreams.

The war diary continued on September 6, with the notation that firing was resumed after a long break on orders from Army Command. Probably they were short of ammunition, for at the end of August Erwin went to Leibach to collect some.

The streets were crowded with an unbroken stream of officers and whores. Most of the female refugees seem to live that way . . . Next day to Oberleibach, a lovely little village. A dimwit of a transport officer. One must extract every hobnail from him with wind from his backside. A moving scene of farewell of a gypsy from his black pony.

On September 19, Erwin wrote:

Interesting events zero. For a long time it was absolutely quiet here, now at night there are front line attacks on the plateau. It is utterly boring. When I have otherwise nothing to do, I fill my mind with the

psychology of the fundamentals of consciousness (memory, associ-
ation, the concept of time).

Tonight again dreamed very vividly of L[otte], a remarkable thing
that. With my own conduct in the matter, the more I review it, the
more satisfied I am with it. Especially that I was so clever at that time
– long ago – to write the letter in question to Johanna [a first cousin
and good friend of Lotte]. Instead of a shackle, it gives me now the
most complete freedom to maintain my dignity, in which I can be
silent as long as I want to. And that was urgently necessary for me. I
make for myself, so to speak, a paper lock, allow nobody to enter,
until the ban is broken in the right way. And then I shall have the
situation that I need, which could never have been created in any
other way. I do not need to speak, because I have already spoken.
Like a coat of mail this one word guards me against every reproach,
ensures me against every disaster from either one side or the other. I
will have only one thing three times for every argument from her side:
peace, peace, peace. Be silent and do not allow yourself to be duped.
Silent as the *earth*, as a buried stone.

What are we to make of this – one of the great intellects of his
time talking to himself about an affair of the heart as he were a
shyster plotting a defense against a lawsuit?

The final entry in the diary is made on September 27:

It is really horrible and I have a home-sickness for work. If this goes
on for long I shall be a wreck physically and mentally. I am no longer
accustomed to work or to think for half an hour. Every rational
thought is entangled with another one: what's the use of it all if the
war is not finally finished. The last two months here were the best of
it. But I have outgrown the whole thing. Is this a life: to sleep, to eat,
and to play cards? Through incessant military service, every bodily
activity is turned into an illusion.

Many days I lie in bed till the midday meal, merely to avoid facing
the question of what I should begin to do in the morning. After
eating, naturally a siesta. To stand up for five minutes is too weary-
ing, so I sit down. I scarcely notice the names of my acquaintances.
Why is it that I am the only one with nothing to do? Is it my fault? Is
it the standard that I apply? Or the special conditions of my demand
for activity – especially because of the futility in which I now live. I
am not completely desperate about it – much too apathetic to be that.
I think: it continues in its idiotic way, nothing can be done. Frightful.

Remarkable: I no longer ask when will the war be over? But: will it
be over? Not naively: hopefully. Are 14 months so horrible? That one
already actually begins to doubt the end.

At this point the war diary breaks off. Schrödinger began to receive some books of philosophy and copies of scientific journals, and he was able to rouse himself from the desperate futility of military life and to turn his mind to the kind of problems that he loved.

MILITARY CITATION

By the end of 1915, a combined German, Bulgarian, and Austrian offensive had finally eliminated Serbia from the war, and some reinforcements could be sent to the Italian front. The third great battle of the Isonzo opened on October 18, with a seventy-hour Italian artillery barrage of unprecedented ferocity. The Fifth Army was in well-fortified positions and when the attack was broken off early in December, the Italian gains had been limited to a small area on the west border of the Karst plateau. The Italians lost 125,000 killed and wounded compared to an Austrian loss of 80,000.

Schrödinger was awarded a citation for his outstanding service during this battle:

In the battle of Oct. 23 to Nov. 13, while acting as a replacement for the battery commander, he commanded the battery with great success. During the preparations as well as in many engagements, he was in command as first officer at the gun emplacement. By his fearlessness and calmness in the face of recurrent heavy enemy artillery fire, he gave to the men a shining example of courage and gallantry. It was owing to his personal presence that the gun emplacement always fulfilled its assignment exactly and with success in the face of heavy enemy fire . . . He has been at the front since July, 1915.

This was entered in his record in the War Archive.

DEATH OF HASENÖHRL

In the Tirolean sector of the Italian front, Fritz Hasenöhrl had been ordered back to duty with a military service cross III Class and his shoulder wound not yet completely healed. At Vielgeruth (now Folgaria) in South Tirol, he was killed by a grenade while leading his battalion of the 14th infantry regiment in an attack. Hermann Mark, then twenty years old, was a corporal in a nearby company, and he recalls the great excitement at the news that the famous physicist had been killed. Hans Thirring later commented that

the moral qualities of duty and sacrifice stand higher than the intellectual qualities of critical thinking and foresight. In those days before Hiroshima there were few intellectuals and researchers who would apply the scientific skepticism and criticism which they used in their work also to the questions of power politics . . . We must have even more respect for his spiritual qualities than we have for his bravery.

Hasenöhrl was posthumously awarded another decoration and the emperor sent his widow a telegram of condolence, which was unprecedented for a mere *Oberleutnant* in the reserves.

On May 1, 1916, Schrödinger was promoted to *Oberleutnant*, and on May 15, after a long winter break, fighting was resumed. The Italians were able to advance about 10 km in a salient around Gorizia and then captured the town. These were the bloodiest of the Isonzo battles, with Italian losses of 286,000 men and the Austro-Hungarian less than half that many. The Italian general Cadorna was impervious to the slaughter of his army for slight tactical gains, and although his grand plan to capture Trieste never seemed likely to succeed, full credit must be given to the tenacious defense by the Imperial Fifth Army.

There is no detailed record of Erwin's military service in this battle, but at about this time he was given charge of a battery which had been installed at Prosecco, about 300 meters above Trieste, and somewhat to the north of the city. He called this a "still more comical naval gun in an extremely boring but beautiful lookout spot." He spent the rest of 1916 here, either by good luck or because, after the death of Hasenöhrl, somebody in Vienna did not want to see his possible successor meet a similar fate.

On November 21, 1916, Franz Josef died, in the eighty-sixth year of his life and the sixty-seventh year of his reign. The last news he heard from the war had been encouraging – the capture of Bucharest. Both he and Kaiser Wilhelm had made

9. Schrödinger as *Leutnant* in the Fortress Artillery,
at the front, 1916

an effort to stop the war in 1916, but the Kaiser's manner was so arrogant that these peace overtures were rejected by the Triple Entente. The new emperor Charles was that rarest of species among the powerful, a sincere Christian. Much to the distress of the military commanders, he forbade any attacks on civilian targets and the use of poison gas. He asked the brother of Empress Zita, Sixtus of Bourbon-Parma, to go to Paris and begin negotiations for peace. Sixtus was an officer in the Belgian Army, since, as a Bourbon, he was legally forbidden to serve in the French armed forces.

Early in 1917, Schrödinger was surprised to see Sixtus making a tour of inspection of the Austrian forces around Trieste. He did not know until later of the peace negotiations. Sixtus persuaded Charles to agree to a separate peace with the Triple Entente, territorial concessions to Italy, and the eventual restoration of Alsace Lorraine to France. As Schrödinger remarked later, this was "a treacherous betrayal of Germany, which unfortunately did not come to pass." The Italians, however, demanded everything promised in the secret Treaty of London, and when the Germans learned of the Sixtus affair, they were able to tighten a virtual noose around the neck of the hapless Charles, who lacked the stamina to resist the military dictatorship that now controlled Germany.

Erwin was now more or less resigned to the continuing war, having recovered from the deadly period of idleness and depression of 1915. He must have been cheered also by a visit from Annemarie Bertel. It is interesting that none of his Vienna girl friends came to see him, but only the country girl from Salzburg, now a young woman aged twenty. According to Anny's account, they did not become lovers at this time. Thus it would appear that Erwin's sexual experience before his marriage was restricted to one love affair, with the probability, however, of casual encounters that he did not record.

METEOROLOGY SERVICE

In the spring of 1917, Schrödinger was transferred back to Vienna. He was assigned to teach an introductory course in meteorology at a school for anti-aircraft officers at Wiener Neustadt (49 kilometers south of Vienna) and also to teach a laboratory course in physics at the University. At this time Hermann Mark first met Schrödinger as his instructor in the physics laboratory. They were both in uniform. As Mark recalled, Schrödinger "was already very impressive to all of us because of his kindness. Of course there was not much physics involved in these experiments that we carried out, but he was infinitely patient with us. – Well that lasted a few months and I had to go back to the front."

Schrödinger's first paper after his return to Vienna was submitted in the last week of July 1917. It was inspired by his experience as an artillery officer, for it deals with the theory of the so-called "outer zone of abnormal audibility" of large explosions. As one moves laterally away from the point of an explosion, the sound level is first attenuated but then may rise again in a zone 50 to 100 kilometers distant, before finally dying out at longer distances. Schrödinger's stint in the army had not dulled his theoretical skills, yet neither had the fallow period led to an outburst of original thinking about problems in the forefront of physics, in particular the problems of quantum theory and atomic structure that were stretching the fabric of classical physics to breaking point. He was still re-acting to various concerns of the somewhat isolated Vienna school, still using his great mathematical facility to make improvements in structures built by others, although he was now thirty years old, an age by which most great theoretical physicists have rebelled against the paradigms received from their university teachers.

About midway through 1917, Schrödinger sent an article on "The Results of New Research on Atomic and Molecular Heats" to *Die Naturwissenschaften* (*The Sciences*). This journal, published by Springer Verlag, was founded and edited by Arnold Berliner, who encouraged authoritative reviews of the status of important fields in addition to brief reports of original research. Schrödinger's paper was a review and did not contain any original ideas. He was apparently unaware of the important work in Berlin of the Danish physical chemist Niels

Bjerrum in 1912, which set forth the quantization of rotational and vibrational energies in molecules. In general the approach to quantum theory in Vienna at this time was through the applications to thermodynamics as made by Planck and Einstein, and not through the spectroscopic studies emphasized by the Copenhagen school of Niels Bohr. Nevertheless, this essay of 1917 marks the first time that Schrödinger wrote anything about the quantum theory, and two years later, it was followed by a much more extensive review of the same general area.

SMOLUCHOWSKI AND STATISTICS

Among Schrödinger's notebooks from 1914–18 are three devoted to the work of Marian Smoluchowski. The notebooks are undated, and it has been suggested that the first one, entitled "Fluctuation Opalescence," may have been written late in 1914, but other historians place it a couple of years later. The other two, entitled "Discussion of the Last Works of Smoluchowski," were written late in 1917 or early in 1918. They deal with the phenomena of Brownian motion and fluctuations diffusion, and the statistical basis of the Second Law of Thermodynamics.

Marian Smoluchowski was born in Vorderbrühl near Vienna in 1872, and attended the upper-class Theresianum Gymnasium at the same time as Fritz Hasenöhrl, with whom he formed a lifelong friendship. He completed his physics studies at Vienna, worked in Glasgow and Berlin, and in 1913 he became professor of physics at Krakow. Schrödinger did not have an opportunity to know Smoluchowski as a co-worker in Vienna, but he came indirectly under his influence through Hasenöhrl, and came to recognize him as the Vienna physicist who was the most noteworthy successor to the heritage of Boltzmann. Smoluchowski died in an epidemic of dysentery in 1917. He is especially remembered for his theory of Brownian motion.

In 1827, the Scottish botanist Robert Brown observed under his microscope a curious perpetual motion of pollen grains suspended in water. "These motions were such as to satisfy me

. . . that they rose neither from currents in the fluid, nor from its gradual evaporation, but belonged to the particle itself." In 1888, Georges Goüy explained how the particles are propelled by collisions with the rapidly moving molecules of the suspension fluid. In the phenomena of Brownian motion, we can see with our own eyes events occurring at the borderline between the macroscopic and the molecular worlds. The perpetual Brownian motion does not contradict the First Law of Thermodynamics, for the source of energy that moves the particles in the kinetic energy of the molecules surrounding them. We may assume that in any region where the microscopic particles gain kinetic energy, there is a corresponding loss in energy of the surrounding molecules, which thus undergo a localized cooling. Brownian motion thus reveals that the Second Law of Thermodynamics is a statistical law: in sufficiently small regions, the entropy does not remain absolutely constant but fluctuates about a mean equilibrium value. In 1905–6 Einstein and Smoluchowski independently provided mathematical theories of Brownian motion. Schrödinger recognized the importance of fluctuation phenomena and plunged into a detailed study of this field, mastering all the papers, and then making original and important contributions to its development. Fluctuations would become one of the leit-motifs of his future scientific work.

Smoluchowski was not troubled by the apparent paradox that natural processes are irreversible, always increasing the entropy of the universe, whereas molecular processes are essentially reversible, and he concluded that "irreversibility is only a subjective concept of the observer, the applicability of which does not depend upon the type of natural process, but rather upon the position of the initial point and the duration of the observation . . . Processes will appear to us to be irreversible, if their initial points lie far beyond the average range of a fluctuation, and if they are observed only for a period of time that is short compared to the time of recurrence." This is a controversial view of irreversibility, which is not accepted by all students of the problem.

Schrödinger's first paper inspired by Smoluchowski was the

solution of the diffusion equation in the presence of a gravitational field, which he completed while on active service at Komaron. Smoluchowski independently solved this same problem and published it at almost the same time, but this coincidence did not bother Schrödinger; he was in fact almost delighted that his result was confirmed by the acknowledged master of the subject. The equation solved by Schrödinger and Smoluchowski was an example of a general type applied by Adriaan Fokker (brother of the aircraft designer) and Max Planck to the time development of stochastic processes (a stochastic variable is one that does not depend in a completely definite way upon the independent variable [e.g., the time t] but is subject to random effects that can only be defined in statistical terms).

PROBLEMS IN STATISTICAL DYNAMICS

During 1918, Schrödinger wrote two long papers on statistical dynamics, devoted to a complete analysis of the random fluctuations in the rate of radioactive decay, the so-called *Schweidler fluctuations*. We are today so familiar with radioactivity that it is impossible to recapture the wonder and mystery that surrounded its manifestations at the time of its discovery. The idea that an atom of an element may exist for a period of time and then suddenly emit a particle and change into an atom of a different element was in itself strange enough, but the fact that the process cannot be influenced by any external factors and appears to be subject to the laws of pure chance was even more remarkable. The best criterion for the randomness of the decay process is the occurrence of fluctuations in the rate of decay that follow the laws of probability theory when measured over successive short time intervals. In a paper presented at the Academy Meeting on March 14, 1918, Schrödinger gave a statistical analysis of such measurements, based on the Fokker-Planck equation in its general form. From this equation, he could obtain the generalized diffusion equations with external forces, for example, the case with elastic forces treated by Smoluchowski, and the case with gravitational forces that he and Smoluchowski had considered earlier. One might view this paper as a resharpening of his theoretical tools after the wasted years in military service. One sees again his deep immersion in the subjects cultivated by the older generation of Vienna physicists, and the choice of problems governed by a reaction to local, almost parochial, interests. Even if the subject matter is classical and unexciting, one must admire his ability to write papers of such elegance while he was undernourished and cold and beset by grievous family troubles.

Schrödinger's second paper on the statistical dynamics of radioactive decay was entitled "Studies in Probability Theory of Schweidler Fluctuations, especially the Theory of their Measurement." It was presented before the Academy on January 16, 1919, and has the distinction of being the longest paper he ever wrote, occupying sixty pages in the *Proceedings*.

GENERAL RELATIVITY

Schrödinger first learned of Einstein's theory of general relativity while he was stationed at the front of Prosecco. He immediately recognized the importance of this theory, which by representing the gravitational field as a consequence of the geometry of the universe, opened new vistas for the world view of physics. When he returned to Vienna in 1917, he found that the university physicists were also excited by the Einstein work. Ludwig Flamm, who had married Boltzmann's youngest daughter Elsa, had already published a paper in the field, and Hans Thirring was working on another application. For the first time in his scientific life, Erwin was able to enjoy critical discussions with knowledgeable colleagues about a subject at the forefront of contemporary physics. His own intensive studies of the mathematical foundations of general relativity are recorded in three notebooks (undated as usual) entitled *Tensoranalytische Mechanik*. Notebook III also included an outline of the analogies between mechanics and optics, such as the relation between Huyghens's principle and Hamilton's equations, a subject that was to play a leading role in his development of wave mechanics in 1925.

The short paper that he sent to *Physikalische Zeitschrift* in November 1917, went to the heart of a fundamental question, how to express the total energy and momentum of a closed system in terms of the formulas of general relativity. In 1916, Einstein had introduced an entity which he called the energy components of the gravitational field. This entity, however, is not a general tensor density and hence is not invariant with respect to transformation of the co-ordinate system. Schrödinger considered a special system for which solutions to the general-relativity equations had been obtained by Karl Schwarzschild in 1916, the surroundings of a stationary sphere of incompressible gravitating fluid. He calculated the sixteen energy components for a co-ordinate system "almost equiv-alent" to a Cartesian one, and found the strange result that they all vanished. The problem raised by Schrödinger is part of a complex of questions relating to the localization of gravi-tational energy, which continue to be studied even today. It is remarkable that in his first note on general relativity he was able to uncover such a deep problem, but it is typical of an approach that became increasingly evident in his work, a disinterest in facile applications and a concern for fundamental principles, often in frontier areas where theoretical physics meets metaphysics.

In a second note, his starting point was a recent paper by Einstein which gave a system of energy components and gravi-tational potentials that exactly integrated the field equations and provided an approximation to a possible large-scale struc-ture of space and the distribution of matter in it. The model consisted of a resting fluid of uniform density which forms a closed space of finite volume having the metrical properties of hypersphere. Einstein had included a "cosmological constant" in his solution, but Schrödinger showed that there was another solution without this constant. His solution had some curious properties: the fluid universe was under a considerable tension and its mass density was zero. The latter property was explained away by a Machian suggestion that non-vanishing mass arises only in the form of mass differences in a non-uniform universe. Einstein replied immediately to this note, saying "The path taken by Schrödinger does not seem to me to be an accessible one, since it leads too deeply into a thicket of hypotheses." Nevertheless this note is interesting since it is Schrödinger's first essay in cosmology, and nineteen years later, after he had studied the work of Eddington, he added a handwritten footnote to calculate the value of the tension on this model universe.

THE END OF THE WAR

During 1917, both Kaisers, Charles and Wilhelm, as well as Lloyd George in Britain, tried to bring about negotiations for peace, but by now the generals were in full command of the warring nations. From their comfortable headquarters, they lusted for the one final offensive that would complete the slaughter of their enemies. Hindenburg and Ludendorff assured Wilhelm that unrestricted submarine warfare would bring Britain to its knees within a few months. Charles warned against this suicidal policy, but Wilhelm dragged him, still pleading for peace, into the final debacle. On April 6, 1917, the United States declared war on Germany – it did not declare war on Austria-Hungary till eight months later.

The multi-national Austro-Hungarian Army had fought courageously for four years. Of the eight million men mobilized, more than one million were killed and almost two million taken prisoners. The generals on both sides of the war were devoted to the doctrine of attack, which consisted in sending masses of soldiers against machine-gun emplacements. When Pershing arrived with the American forces in France, he delegated one man in ten to the military police, whose duty it was to follow the attackers and shoot any laggards.

The supply situation in Austria-Hungary was deteriorating. The soldiers lived mostly on soup made of dried vegetables. The meat ration was 200 grams a week in the front line and 100 grams in the rear echelons, less than a McDonald hamburger, but usually lean horse meat. The meat was often full of worms, but the government informed the troops that, though unappetizing, they were not dangerous to health. In some divisions, only front line troops had uniforms; the reserves waited in underwear until it was their turn to be thrown into the slaughter. Under these conditions, Conrad persauded Charles to launch a "final offensive" on the Italian Front; when it foundered with heavy losses, he was finally retired.

Schrödinger remained on active service till the end of 1918. Among his random jottings from that year a curious fragment has been preserved:

Da sind zwei Armeen ausmarschiert seit 1914
Die eine kämpft noch
Die von der andern haben Frieden gemacht (unter) der Erde
Wahle! Zu welcher willst du?

Two armies have marched out since 1914 / One is fighting still / Those of the other have made an underground peace / Choose! Which one do you want to belong to?

This may have been written by Erwin himself or copied from some illegal peace propaganda, which was beginning to appear more openly. Natalie Lechner-Bauer became so active in the peace movement that she was arrested and convicted of treason.

During these times, life was particularly difficult in the large fifth-floor apartment which the Schrödingers rented from grandfather Bauer. They had never installed electric lighting, partly because Alexander did not wish to pay for it, and partly because Rudolf preferred the mellow white light of the Welsbach gas mantles to the reddish-yellow of the available electric bulbs. They had removed the great tiled stoves from several rooms and replaced them with gas fires. They also cooked with gas, although a huge slow-combustion stove still stood like a monster in the middle of the kitchen. Thus when word came from City Hall that every dwelling was to be allowed only one cubic meter of gas per day – any violation to be punished by complete cut-off – the Schrödinger household had serious energy problems in addition to the terrible shortage of food. Erwin's mother appeared to be recovering from a radical operation for breast cancer, which had been performed in 1917, but she was still weak and in pain. Erwin's own health was far from adequate: in August 1918, an inflammation of the apex of the lung was diagnosed, which may have been tuberculosis, since this disease was epidemic among the undernourished population of the city.

Rudolf's business had been destroyed by wartime shortages, and in 1917 Groll Brothers, his oilcloth and linoleum company on Stephansplatz, was closed. When the war ended, he no longer had either the money or the strength to try to salvage it. Thus the family for the first time was facing serious financial difficulties, since Erwin's military pay stopped at the end of 1918 and he had barely enough income from the university to support himself. During this winter the family often ate at a community soup kitchen.

After the Armistice, as the Empire disintegrated, the situation in Vienna became much worse than during the actual war, for food supplies from Hungary were cut off, and the Triple Entente maintained its blockade. Thousands of people in Vienna were starving and freezing in the winter of 1918–19. The streets were filled with beggars, ex-soldiers exhibiting various mutilations, with decorations pinned to their rags. The worst were the *Zitterer* (tremblers) who suffered grotesque facial tics and continuous jerky movements.

The women had most of the responsibility for foraging for their families; the farmers in outlying areas had large stocks of food, but it was necessary to seek them out and barter precious possessions for a few eggs, vegetables, and other edibles. Even to obtain these, ladies had to importune and beg in ways that they found humiliating, and the farmers drove hard bargains with the starving city dwellers.

Schrödinger recalled that "In 1918, we had a sort of revolution; Emperor Karl went and we became a republic, without changing much in our lives. For me personally the dismemberment of the country had one result, since I had a call as Extraordinarius to Czernowitz, where I intended to lecture on theoretical physics, but in private life to concern myself more with philosophy (I was just now with great enthusiasm becoming familiar with Schopenhauer and, through him, with the doctrine of unity taught by the Upanishads)." Czernowitz, however, had become a part of Romania and the new government would not tolerate foreign teachers.

The Austrian government tried to make the country part of Germany, but this action was blocked by France. Italian troops occupied Vienna and the Czechs cut off the export of coal. Many lives were saved by the smuggler (*Schleichhändler*) who sneaked across the border from Hungary and called once a week with food supplies; paper money was useless, but everything else from carpets to grand pianos was carted away. Karl Renner, the socialist president, implored the British and French to raise the blockade so as to allow some food to enter the country. The blockaded Germans sent a little dried fish, Sweden sent rice, and the Swiss and Italians did what they could to provide some relief. On Christmas Day, the official toll of war dead on both sides was announced as seven and a half million.

Herbert Hoover came to Vienna on January 17, and arranged for some American relief through Quaker agencies. His motives were as much political as humanitarian, and food was used as a weapon to stop the spread of bolshevism. In March, as the blockade continued, it became a crime to forage in the countryside around Vienna. Finally, on March 22, the blockade of Austria was lifted, although it continued to be applied to Germany. It is difficult to find a parallel in history to this deliberate starvation of a defeated enemy. On March 23, the emperor left the country, to die in exile after one final typically inept attempt to regain his throne.

CHAPTER 4

From Vienna to Zurich

In the midst of all the post-war turmoil and suffering, Schrödinger took no respite from his intensive research at the Physics Institute. He also filled notebook after notebook with notes and commentaries based upon this reading of European and Eastern philosophers. It was in these dying days of the Danube Empire that he formed the foundations of his philosophy, which was to remain remarkably constant all his life.

SCHOPENHAUER

Schrödinger read everything written by Schopenhauer. The luminous words of the philosopher of pessimism were perfectly suited to the world of 1919, and in that context may even have had a consoling effect, making some sense of Europe's four years of self destruction.

Arthur Schopenhauer (1788–1860) was born in Danzig, the son of a rich Hanseatic merchant and a romantic novelist. As a youth, he traveled widely, becoming fluent in English and French, so that his prose acquired a clarity quite unlike the murky philosophical German of his times. His first education was that of a man of the world, only later did he acquire the usual academic credentials.

He became a friend of Goethe, and in 1816 wrote a small book on color theory inspired by the ideas of the older man. Schopenhauer regarded himself as the true spiritual descendant of Kant, and he did not hide his view that Hegel, then at the height of his fame, and other university philosophers were charlatans who had perverted the Kantian gospels.

The first edition of his major work, *The World as Will and Representation* appeared in 1818. He followed Kant in the belief that the mind is not merely a passive recipient of sense impressions, but takes an active role in fitting phenomena into the categories of space and time, the principle of causality being the necessary method for creating his representation of the world. Kant taught that the real world, the *noumenon*, the thing-in-itself (*Ding an sich*), can never be accessible to human thought or experience. Schopenhauer did not agree: he believed that the thing-in-itself can be identified as *will*. Every person experiences himself in two different ways, as an object like any other, and through self-consciousness as a will. The will is neither a phenomenon nor a representation, it is a directly experienced reality. What is true of man, is also true of the world: its thing-in-itself is will. On the foundation of this primary intuition, which of course can be neither proved nor disproved, Schopenhauer constructed a philosophy that has continued to fascinate and influence thinkers of all kinds: Wagner, Nietzsche, Thomas Mann, Freud, Klimt, and Schrödinger, to mention but a few.

Schopenhauer kept a faithful dog called "Atman" (Sanskrit: the soul), a copy of the Hindu scriptures at his bedside, and an ancient statue of the Buddha dressed as a beggar and thickly covered in gold leaf. His philosophy is closely related to the ancient wisdom of the East, and many westerners first learned of Vedanta and the Upanishads through his writings. His direct influence on Schrödinger was considerable, but equally important was the introduction he provided to Indian philosophy. Schopenhauer often called himself an atheist, as did Schrödinger, and if Buddhism and Vedanta can be truly described as atheistic religions, both the philosopher and his scientific disciple were indeed atheists. They both rejected the idea of a "personal God," and Schopenhauer thought that "pantheism is only a euphemism for atheism." Yet Schopenhauer claimed kinship not only with the Buddha but also with the Christian mystics, and although Meister Eckhart may have been a heretic, he was not an atheist.

VEDANTA

Schrödinger read widely and thought deeply about the teachings of the Hindu scriptures, reworked them into his own words, and ultimately came to believe in them. Possibly his half-famished state at this time was an involuntary mortification of the flesh conducive to religious experience.

In July 1918, he wrote:

Nirvana is a state of pure blissful knowledge . . . It has nothing to do with the individual. The ego or its separation is an illusion. Indeed in a certain sense two "I's" are identical, namely, when one disregards all special content – their *Karma*. The goal of man is to preserve his Karma and to develop it further. The goal of woman is similar but somewhat different: namely, so to speak, to create an abode that accepts the Karma of man . . . When a man dies, his Karma lives and creates for itself another carrier.

It would appear that Erwin was unwilling to give up his Ego completely, or even his male supremacy, at least at this stage in his studies.

In August he wrote: "The stages of human development are to strive for: (1) *Besitz* (2) *Wissen* (3) *Können* (4) *Sein*. [(1) Possession (2) Knowledge (3) Ability (4) Being.]"

Schrödinger wrote that he was "under the very strong influence" of Lafcadio Hearn (1850–1904). Hearn was born in the Ionian Isles of Irish-Greek parentage, went to America at the age of nineteen, and from the age of forty lived in Japan, where he immersed himself in Japanese Buddhist culture. In his essay on "The Diamond Cutter," he wrote:

The ego is only an aggregate of countless illusions, a phantom shell, a bubble sure to break. It is *Karma*. Acts and thoughts are forces integrating themselves into material and mental phenomena – into what we call objective and subjective appearances . . . The universe is the integration of acts and thoughts. Even swords and things of metal are manifestations of spirit. There is no birth and death but the birth and death of Karma in some form or condition. There is one reality but there is no permanent individual.

Phantom succeeds to phantom, as undulations to undulations over the ghostly Sea of Birth and Death. And even as the storming of a sea is a motion of undulation, not of translation, – even as it is the form of the wave only, not the wave itself, that travels – so in the passing of lives there is only the rising and vanishing of forms – forms mental, forms material. The fathomless Reality does not pass . . . Within every creature incarnate sleeps the Infinite Intelligence unevolved, hidden, unfelt, unknown – yet destined from all eternities to waken at last, to rend away the ghostly web of sensuous mind, to break forever its chrysalis of flesh, and pass to the extreme conquest of Space and Time.

Perhaps these thoughts recurred to Erwin when he made his great discovery of wave mechanics and found the reality of physics in wave motions, and also when he based this reality on an underlying unity of mind. Yet, in the course of this life, belief in Vedanta remained strangely dissociated from both his interpretation of scientific work and his relations with other persons. He did not achieve a true integration of his beliefs with his actions. The *Bhagavad-Gita* teaches that there are three paths to salvation: the path of devotion, the path of works, and the path of knowledge. By inborn temperament and by early nurture Schrödinger was destined to follow the last of these paths. His intellect showed him the way, and throughout his life he expressed in graceful essays his belief in Vedanta, but he remained what the Indians call a *Mahavit*, a person who knows the theory but has failed to achieve a practical realization of it in his own life. From the *Chandogya Upanishad*: "I am a Mahavit, a knower of the word, and not an Atmavit, a knower of Atman."

DEATH OF THE FATHER

The first post-war years were filled with difficulties and sorrows for Erwin, the loss of father, mother, and grandfather within a span of less than twenty-one months. In the summer of 1919, the family was able to afford a lakeside holiday at Millstatt. While there, Rudolf, then sixty-two years old, showed the first signs of the hypertension and atherosclerosis that already must have been advanced. On the walks and excursions which he always loved, he fell behind the others and had difficulties in climbing hills. He tried to hid his debility by saying that he was

collecting botanical specimens for his research on morpho-
genetics. When they returned to Vienna, there were more
ominous symptoms, nosebleeds, retinal hemorrhages, and
oedema. The climb to their fifth-storey apartment became an
ordeal for him. With the severe gas shortages, they faced both
cold and darkness. He bought some carbide mine lamps to
illuminate his library, but they pervaded the whole house with
a dreadful smell.

Rudolf was the only one truly aware of their precarious
financial situation. With the failure of his business, they were
dependent upon income from investments, and many of these
were in government bonds. Late in 1919, the cost of living
began to rise steeply, although rampant inflation did not begin
until about two years later. Rudolf was worried that Erwin had
practically no income, only 2,000 crowns that year from the
university, at a time when about 2,000 crowns a month was
needed to provide for an average family. Nevertheless, the
university was a refuge from the troubles at home, and Erwin
spent every day there, managing to accomplish a remarkable
amount of work under the most difficult circumstances.

Rudolf's strength declined steadily, and he died on
December 24, peacefully resting in a grandfather chair. Erwin
later reproached himself for not securing better medical care
for his father. In 1930, he published a short contribution in the
Christmas Eve issue of a German paper, under the heading
"Neglected duties":

Thus there was no longer much more to destroy in the sentimental
complex of Christmas eve, when, on 24 December 1919, between
6 and 7 in the evening, my father died. And strange. Also on this
evening, a beloved person, who knew nothing of the change for the
worse in the last days, was not with me but in the circle of her family.
From there, sent with loving, unsuspecting intentions, an hour later a
messenger with a basket of presents stood at the door. – With all that,
it served me right. For however reluctantly I confess it to myself
today: had the immediate crisis not come, I would also on this
evening, as so often in the previous weeks, have left my mother alone
for several hours with the desperately sick man; and this, although I
could understand that the holy day meant something to her, and that
my father would never experience another one. So to me Christmas is

a feast which I am not fond of, from which I expect nothing good, and which more than any other reminds me of neglected duties.

The Christmas hamper had probably been sent by Anny from Salzburg. During the summer, Erwin had been able to see her quite often, since Millstatt is only about 120 kilometers from Salzburg. During the autumn, he made a special visit during which they became lovers and also became engaged. Anny came to Vienna and obtained a position as secretary to General Direktor Fritz Bauer of the Phoenix Insurance Company. The Bauers were a wealthy Jewish family (not related to Erwin's grandfather). Anny's monthly salary was more than Erwin's annual income, a situation that caused him embarrassment and bitterness. He used to spend almost every evening with her, but later, in a rather irrational way, he reproached her for keeping him away from his dying father.

Anny once took her fiancé to visit the Bauers in their impressive home. He met the family, including fourteen-year-old Hansi, who found him to be "stiff and suffering" during afternoon tea.

For quite a time after his father's death Erwin had a recurrent dream that he had recklessly sold off all his father's scientific instruments and books while his father was still alive: "I believe this anxiety dream was caused by a bad conscience due to my less than honorable conduct towards my parents in the years 1919/21. This explanation is consistent with the fact that I have usually been exempt from nightmares because of course I very seldom, as they say, 'have a bad conscience about anything'."

COHERENCE EXPERIMENTS

To do any scientific research in Vienna in 1919 must have required unusual powers of concentration. Early in the year Erwin decided to spend some time on an experimental project. He always enjoyed precise optical measurements, and the problem he chose required interferometric techniques, which illustrate so beautifully the wave properties of light. The immediate motivation for his work was a recent paper by

Einstein which discussed the emission of light as particles (photons). Schrödinger found it difficult to reconcile this theory with the well-known wave properties of light. He had the idea that a suitable experimental test should be able to decide between particles and waves. Thus if an excited atom emits a photon, the emitted light should travel along a path of narrowly defined angular width, whereas if the atom emits a wave the light should constitute a spherical wavefront spreading uniformly in all directions from the point source. For Schrödinger, and for almost all other physicists in 1919, it seemed obvious that these two pictures could not both be true. He thought at first that if the particle theory is true, the degree of coherence between two "light bundles" emitted from a small source should be less at large than at small angles of separation, and that such an effect might be detected by examining the interference between the two bundles. No sooner had he started to test this idea experimentally than he realized it was invalid, since in the interferometer the two light sources are defined by two slits, from which the light spreads out in all directions. His experimental approach had been based on the use of extremely small light sources, and he also soon realized that even the smallest laboratory source is enormous compared to the dimensions of an atomic or molecular emitter of light. He persisted anyway, simply to study the effect of source dimensions on the clarity of the interference fringes.

His light source was an electrically heated filament, thinner than had ever been used before, 2–4 μm platinum wire. This could not be used in a vacuum lamp, but it was possible to maintain a small length, about 1 millimeter, at red heat in air for long enough to make his observations, provided he worked quickly. With this arrangement, he was able to detect interference fringes at angular ray separations of up to 56°, much larger than ever seen previously. He concluded that he had clearly demonstrated the increase in the limiting angle for well-defined interference (and hence coherence) as the diameter of the light source is decreased, which is the result he expected from a purely mathematical analysis of the problem.

This research was published in the *Annalen der Physik*, the most prestigious German physics journal. Except for a few observations on color vision, it was his last work as an experimental physicist, and it demonstrated laboratory skill of a remarkable delicacy and precision. The long hours spent sitting in front of his microscope in the dark, cold laboratory, watching the elusive interference fringes come and go, must have left him with a firm conviction that light waves have an incontrovertible physical reality. This research was also important for his future in that it made him study and ponder the strange ambiguities of the quantum theory of light.

COLOR THEORY

Schrödinger's most important research at the University of Vienna from 1918 to 1920 was in the field of color theory (*Farbenlehre*). He continued to publish occasional papers on this subject between 1921 and 1925, but the basic papers appeared in the *Annalen* in 1920. As with most of his work since graduation, the origin of his interest in color theory was research in progress by his colleagues at the University. From 1917 to 1920 his friend Fritz Kohlrausch held a lectureship in color theory at the School of Applied Arts, during which he investigated the colors of artists' pigments in terms of their perceived hue, saturation, and brightness (or lightness). These are the three most distinctive perceptual qualities of color. The term *hue* specifies the spectral color (wavelength) perceptually most similar to an ordinary broad-band color; *saturation* specifies the amount of white that seems to be mixed with a spectral (monochromatic) color in an ordinary color (the less whitish it appears, the more saturated it is); *brightness* is related to the perception of the intensity of light contained in, or reflected by, a color. Erwin's old professor, Franz Exner, also presented experimental papers before the Vienna Academy in 1918 and 1920 on the brightness of artists' colors.

The inspiration of Schrödinger's work on color theory seems to have been the philosophy of Ernst Mach, which he had continued to study intensively during his service with the artillery. Human color vision provides one of the best examples of the Machian elements of sensation. Schrödinger based his color theory on direct observations in which he matched areas of colored light for hue, brightness, and saturation. He divided that subject into "elementary color theory" in which colors are matched in all three perceptual respects, so that the sensations become indistinguishable, and "advanced color theory" in which the sensations differ (e.g., comparison of brightness between patches of different hue). Sensations of color can become extremely complex, and all the theories considered by Schrödinger are based upon comparisons between isolated patches of color, so that effects due to contrast with different backgrounds, temporal after-images, and other factors are all excluded.

Schrödinger's first paper on color was "Theory of Pigments of Maximal Luminosity," which was submitted to *Annalen* just before the Christmas of 1919. It recalls the great interest of grandfather Bauer in artists' pigments, and Erwin would certainly have discussed this paper with the old man.

The color of light that is reflected from a streak of pigment never reaches the degree of saturation of a pure spectral light, but always appears more or less whitish compared to a pure light of the same hue . . . The impossibility of realizing in pigments colors of spectral saturation is not a technical problem but a matter of principle . . . To achieve the full saturation of a spectral color, a pigment can actually reflect only an infinitesimal range of wavelengths while completely absorbing everything else. In this case, as Helmholtz has already remarked, it would appear to be extremely dark, in the limit, black.

Given this limitation, what is the greatest brightness that can be achieved with mixtures of pigments?

Schrödinger introduced the concept of *ideal colors*. These are colors for which the spectral reflectance has values only of either 0 or 1. In practice, of course, the reflectance does not jump abruptly from 0 to 1, but some actual pigments approximate this behavior quite closely. Schrödinger proved that if a color is obtained by mixing pigments having the characteristics of ideal colors, it will have the maximal brightness obtainable for that color. The ideal colors are therefore also *optimal* colors. This paper was an important contribution to color theory and the concept of ideal colors since has been used in the analysis of many color problems.

Schrödinger's next work on color theory was a long three-part paper, submitted to *Annalen* in March 1920, on "Fundamentals of a Theory of Color Measurement in Daylight Vision." This was his major contribution to the subject, and it was honored by the prestigious Haitinger Prize of the Vienna Academy of Sciences. He aimed to present a logical account of the state of the subject as well as the original advances he himself had made. The restriction to daylight vision allowed him to consider color vision independently of light intensity; in terms of physiology it is almost exclusively cone vision.

Schrödinger based his analysis of color vision on the three-color theory of Thomas Young (1806), surely the most prescient work in all of psychophysics, which was rediscovered, developed, and extended by Hermann Helmholtz in the latter part of the nineteenth century. The Young–Helmholtz theory is based on the hypothesis (since proven) that the normal (trichromat) human retina contains three types of receptors, each with a particular spectral response curve; these may be called *red*, *green*, and *blue* receptors on the basis of their spectral responses. Any spectral color (light) or any mixture of such colors can be matched by a linear combination of the three basic colors.

The *Annalen* papers, which have been called "masterly" by modern experts in color theory, are impressive examples, probably the most impressive in all the scientific literature, of how the philosophy of Mach can be applied to an actual problem. The data are not pointer readings or mechanical measurements of any kind, but rather the direct results of human sensations. Yet, despite its profundity and elegance, this work has had little consequence in practical color measurement, which still relies on empirical tables and color charts. In real life, color vision does not involve simply the comparison of isolated patches of color. It may not be possible to fit its complex reality into the framework of Mach's "elements of sensation," since color vision is determined by the total *Gestalt* of visual perception, which does not contain "elements" at all in the Machian sense.

In 1925, he published an article on color in *Naturwissenschaften*, "On the Subjective Colors of the Stars and the Quality of Twilight Sensitivity." The problem was to explain the discrepancy between the colors seen for fixed stars and their temperatures, as compared with the colors and temperatures of light sources on earth. For example a source that appears white on earth appears yellowish for a star of about the same temperature. This paper includes some experimental observations made with a simple device that enables an observer to compare daylight colors (cone vision) with dark adapted colors (rod vision). Rod vision is not completely colorless but has a bluish tinge. Four normal trichromats matched the twilight color as a pale unsaturated violet similar to the color of the common lilac. Thus Schrödinger explained the yellowish appearance of the white star as the consequence of contrast against the lilac background of the rod vision.

In the early 1920s Schrödinger became recognized as the world authority on color theory. Accordingly he was asked by the editor of the new eleventh edition of Muller-Pouillet, *Lehrbuch der Physik*, to write the section on "The Visual Sensations." The *Lehrbuch*, which originally appeared in 1868, was one of the best of those many-volume scientific encyclopedias that have been faithfully produced by German scientists for many generations. They define in authoritative terms the contemporary status of a subject and provide a solid foundation for future advances. Schrödinger produced a 104-page article with hundreds of references, which for many years was the standard work on the subject. Nowadays such articles are prepared with the help of computer searches and skilled assistants, but for him it was a lonely and time-consuming task. The article displays his mastery of not only the mathematics of color measurements, but also the optics and physiology of the eye. Schrödinger was a worthy successor to Helmholtz in this field.

Yet color theory was a field attractive to psychologists and physiologists, but apparently with few connections with basic physical problems. The clue to Erwin's great interest in it was probably that he was still tempted by his wartime vision of himself as a philosopher, still immersed in Mach, Schopenhauer, and by now also in eastern mysticism. Color theory stands at the crux of the ancient mind–body problem. By working in this field Erwin could be both a philosopher and a scientist.

MARRIAGE AND DEPARTURE

In January 1920, the Faculty proposed Schrödinger for an associate professorship in theoretical physics. The salary would have been inadequate to support a wife and he was anxious to marry Anny as soon as possible, and without having to rely on her income as a secretary. He had been offered an assistantship with Max Wien in Jena, with an assignment to give some theoretical lectures. The salary of 2,000 marks per annum appeared to be satisfactory, since the runaway German inflation had not yet begun. Hans Thirring, who had some outside income from consulting, accepted the associate professorship in Vienna and was ultimately (1927) named professor of theoretical physics, after he had virtually ceased any research in that subject. In retrospect, one can say that Schrödinger was fortunate to escape from Vienna when he did, for it is most unlikely that his great work could ever have been accomplished there.

Schrödinger and Annemarie Bertel were married twice, the first time in the rectory of the Catholic Church of St. Leopold, in the parish where Anny was living. The second ceremony was a more formal wedding on April 6 in the Evangelical Church on Martinsgasse. This duplication of the sacrament must have been a good omen for the duration of their marriage, which despite all stresses endured until they were parted by death.

When they were married Anny was twenty-three and Erwin thirty-two, and they were in some ways an incompatible couple. Anny had little formal education and limited intel-

10. Erwin and Annemarie, wedding picture, March 1920

lectual interests. She played the piano not badly and had a pleasant singing voice, but these talents would have little appeal for Erwin, and he never allowed her to have a piano. She was considerate and kind-hearted, with an outgoing personality. She was not extremely feminine, a bit of a tom-boy as a girl, and becoming distinctly mannish in appearance as she grew older. Although Erwin probably did not suspect this at the time, she was either unable to have children, for reasons that are obscure, or was deterred from having any by a history of mental instability in her family. One of her greatest attractions was her fervent admiration for everything about Erwin, his looks, his personality, and his intellectual brilliance. She once told Hans Thirring, "You know it would be easier to live with a canary bird than with a race horse, but I prefer the race horse." She entered the marriage with the hope and expectation that it would be a true union of minds and bodies, in which she would achieve happiness through boundless submission to her brilliant and beautiful lover. These illusions may have lasted for at least a year.

Considering the frequency of his post-marital affairs, it is noteworthy that Erwin did not record any serious loves between the time he was dismissed by Felicie and the time he became engaged to Anny. During these eight years, he must have had casual sexual encounters, with sweet Vienna maids and the like, and during the war both professional and amateur sex were readily available. After the war, he had no home of his own and the appeal of domesticity must have been considerable. In this regard, Anny became an excellent wife, relieving him of all everyday concerns, providing the kinds of food and wine that he preferred, nursing him when he was ill, and after his interest in her as a sexual partner disappeared, remaining his friend and even helping him to find other feminine companionship. From his viewpoint, it was hardly an ideal marriage, but it had many compensations. In her case, it eventually led to emotional frustration, only partly compensated by her satisfaction in being the wife of a great man. A neutral observer must find them to be a strange couple, surrounded by a certain mystery, which could only be solved by a much more intimate understanding of their psychosexual histories.

JENA, STUTTGART, BRESLAU

Erwin and his bride arrived in Jena in April. It was a charming small city of about 70,000 nestled in the valley of the River Saale. The university, founded in the middle of the sixteenth century, had enjoyed its greatest fame from 1790 to 1805, when Friedrich Schiller was professor of history and his friend Goethe lived only a short distance away at the petty court of the Duke of Weimar. More recently it had become an important center for scientific research, gaining support from the large Zeiss Optical works established nearby.

The Schrödingers received a hospitable welcome from Max Wien and his wife, whom, however, they found to be "anti-semites, but more as a matter of custom, not very bad." Max

Wien's brother Wilhelm (Willy) was one of the most powerful figures in German physics, professor at Munich, editor of *Annalen*, and a Nobel laureate in 1911 for his discovery of a semi-empirical "black-body" radiation law.

Schrödinger immediately made a favorable impression on the physics community in Jena, both in the laboratory and in his lecturing, for he had been there only a few weeks when the faculty recommended his promotion to an associate professorship. The German inflation was worsening, and even though his total income in Jena would now amount to about 11,000 marks a year, his associate professorship was not a permanent position. Thus when a tenured associate professorship was offered at the Technische Hochschule in Stuttgart, he resigned from his various posts at Jena as of October 1, and moved with his wife to the great industrial city and capital of Baden-Württemberg.

The professor of experimental physics in Stuttgart was Erich Regener, an expert in X-ray physics, who had worked for many years in Berlin with colleagues including Otto Hahn, Lise Meitner, Gustav Hertz, and Albert Einstein. While in Stuttgart, Erwin also often met Hans Reichenbach, who was lecturing in mathematics but even then had a great interest in the philosophy of science. Although he had never studied in Vienna, he became an ardent exponent of logical positivism and later wrote several books on the philosophy of quantum mechanics.

The closest friends of the Schrödingers in Stuttgart were the Ewalds, Paul and Ella. Paul was appointed professor of theoretical physics in 1921 and arrived early that year, but Ella stayed in Munich until her second son Arnold, named after Sommerfeld, was born. Paul was a big, handsome man, and Anny was greatly attracted to him, while perhaps responding to her unfeigned admiration, he in return became fond of her.

While Erwin and Anny were in Stuttgart, his mother became seriously ill with a recurrence of cancer. Fortunately she had been able to visit them while still able to travel, and Anny accompanied her back to Vienna at the end of February. Her condition worsened, and during the three months before her death in September 1921, she was unable to leave her bed. After the death of Rudolf, she had been forced to give up her beautiful Vienna home, since her father needed more rent money to pay the heavy post-war real estate taxes; he was in dire financial straits himself as inflation had destroyed his pension and savings. Thus the widow was turned out and the flat rented to a "rich insurance Jew" – this was the rude phrase used by Schrödinger many years later in recalling the event. It reflects the bitterness he felt at his mother's fate, his frustration that he was unable to do anything about it, and his bad conscience at leaving her in the lurch. (Presumably, however, her two sisters who were comfortably situated were able to take care of her.) Erwin's almost neurotic concern for financial security and widow's pensions was caused by these years of financial destitution, during which he had to leave his native land to secure even a marginally adequate income but was still destitute of assets that could be used to help his mother. In later years he worried continually about what might happen to Anny as a widow, as he was haunted by the vision of his dispossessed mother. As it turned out, during her few years of widowhood, Anny would be more wealthy than at any other time in her life.

In the spring of 1921, several German universities, Kiel, Hamburg, Breslau, and also Vienna, were seeking professors of theoretical physics. Schrödinger was being seriously considered for all these posts. The position at Breslau was the most attractive, giving due weight to both the reputation of the university and the level of the appointment. Otto Lummer, famous for the measurements of black radiation that inspired the Planck quantum hypothesis, was professor of experimental physics there; Fritz Reiche was working on quantum theory, and Rudolf Ladenburg, an outstanding spectroscopist, was also deeply interested in this subject.

Schrödinger had serious doubts about a move to Breslau, since the city was in the heart of the Silesian industrial region near the new Polish border. He wrote to his good friend Stefan Meyer in Vienna that he felt a "certain dread of the mad leftists . . . who are continually gaining more adherents among the workers, and where that can lead has been seen in Halle, where two-thirds of the socialist party have just declared for a union with Moscow, i.e., for the red terror. I often have the feeling: let me get away from this powder keg." He says that he would not mind going alone to Breslau but even the remote possibility of exposing a wife, "and perhaps children," to such circumstances without being able to protect them is the most dreadful thing that can be imagined. In late October, he wrote to Hans Thirring: "Although now as before I aspire to Vienna, it is clear that it would be for me a really embarrassing thing . . . to refuse the Breslau *Ordinarius* for the Vienna *Extraordinarius*." He went on to say that he was not confident that he would be promoted to the Vienna *Ordinarius* when it became available since during recent months he had been passed over by the Austrian government for two full professorships in Graz, even though he had been ranked *primo loco* for both appointments. Erwin did not have political influence with either the clericals or the socialists and thus was at a serious disadvantage in the Austrian intrigues for academic preferment. Thus, despite his misgivings, he accepted the Breslau professorship for the beginning of the summer semester and once again Anny packed their few possessions for another move to a new city, the third within eighteen months.

OLD QUANTUM THEORY — PENETRATING ORBITS

While in Stuttgart, Schrödinger studied intensively the recent book of Arnold Sommerfeld, *Atomic Structure and Spectral Lines*, the first edition of which had been published in 1919. The book had immediately become the "Bible" of its rapidly developing field. It is remarkable that Schrödinger, who was new to the subject, was at once able to make an important contribution, which he sent to *Zeitschrift für Physik* in January 1921. As he explained some years later:

In my scientific work (and moreover also in my life) I have never followed one main line, one program defining a direction for a long time. Although I can work only poorly in collaboration, and unfortunately also not with pupils, my work in this respect is still not entirely independent, since if I am to have an interest in a question, others must also have one. My word is seldom the first, but often the second, and may be inspired by a desire to contradict or to correct, but the consequent extension may turn out to be more important than the correction, which served only as a connection.

Thus, the quantum theory of atomic structure and spectra was already well advanced when Schrödinger wrote his 1921 paper. The interpretation of atomic spectra had been achieved by a young Dane, Niels Bohr, who was working with Rutherford in Manchester. Bohr was only two years older than Schrödinger, and he had made his great discovery in 1913 at the age of twenty-eight. He based his theory on the Rutherford model of the atom, a central positive nucleus surrounded by negative electrons revolving in orbits like planets around a sun. He postulated that in the absorption or emission of radiation corresponding to a spectral line a single electron jumps between two different orbits. When an electron makes a transition between two states with energies E_1 and E_2, the frequency of the spectral line v is given by the Planck–Einstein relation,

$$hv = E_1 - E_2.$$

Sommerfeld extended the Bohr theory to consider not only circular electron orbits but also elliptical ones.

Schrödinger pointed out that elliptical and circular orbits similar to those proposed for the electron in the hydrogen atom cannot occur in an atom like sodium. The sodium atom has eleven electrons, two in the innermost shell, eight in the next shell, and a sole electron in the outermost shell. The supposed elliptical orbit of the outer electron must sometimes penetrate the shielding shell of eight and experience an effective nuclear charge closer to $Z' = 9$ than to $Z = 1$, and thus the suggested elliptical orbit based on $Z = 1$ is unrealistic. The exact calculation of the true orbit "appears to be hopelessly complicated," but he made an approximate calculation, based on a model in which the electron passes smoothly from a large outer elliptical orbit to a small inner one as it penetrates the shielding shell. The concept of penetration of outer electrons through a shielding shell of inner electrons was frequently used in later work on atomic structure.

THE UNIVERSITY OF ZURICH

Erwin and Anny had only just returned to Breslau after witnessing the interment of his mother in Vienna's Hietzinger Friedhof, when they prepared to leave turbulent Germany for peaceful Switzerland, in response to a call from the University of Zurich.

The University of Zurich was founded in 1833. It was not organized from above, but in response to demands by the people of the district for an institution of higher learning. Even by 1860, Zurich was still a small town with about 20,000 inhabitants, but it grew rapidly, to 200,000 in 1920. In 1855, to meet the need for technical education, the Eidgenossische Technische Hochschule (always called the E.T.H., ay tay ha) was opened, and Rudolf Clausius was named professor of physics there. In 1857 he became also professor at the University. Clausius was one of the greatest physicists of the nineteenth century. He established the Second Law of Thermodynamics and defined the concept of entropy; he also made important contributions to the kinetic theory of gases. In 1878, Alfred Kleiner was named as associate professor of experimental physics. Kleiner's greatest work, as he himself often admitted, was his approval in 1905 of Albert Einstein's thesis, and his influence in calling Einstein in 1909 to a post as *Extraordinarius* in Theoretical Physics, the first appointment in this subject at the university. In 1922 Einstein went to Prague as full professor; he returned to Zurich in 1912, but soon moved to the Kaiser Wilhelm Institute in Berlin, a city that was then the center of world physics.

The Swiss authorities tended to be parsimonious; while the university was able to attract excellent young physicists, they were easily lured away by higher salaries and better research conditions elsewhere. In 1911, the brilliant Dutch chemical physicist Pieter Debye, who had been working with Sommerfeld in Munich, was appointed to the Einstein vacancy, but only as *Extraordinarius*. He did his important work on the theory of heat capacity of solids while at Zurich. When he received an offer from Tübingen, he was promoted to the first full professorship of physics at Zurich in 1912, but left soon after to go to Utrecht. To succeed Debye, another brilliant choice was made, Max Laue, also from Sommerfeld's department. He had already made his great discovery of X-ray diffraction and was to receive the Nobel prize a few years later, but the frugal Swiss gave him an associate professorship. Following the now familiar pattern, he left after two years for a professorship at Frankfurt. An attempt was made to get Debye back, but the salary offered was not enough. In 1920, however, the E.T.H. had more success, and he stayed there for seven years before moving to Leipzig.

This musical chairs at the German-language universities was an accepted game. The curricula at all these universities were much the same and there was a well-established order of prestige, with Berlin, Munich, and Göttingen at the head and places like Aachen and Innsbruck at the foot. If he received an offer of a full professorship at another university, an associate professor was expected to transfer even if the new place was lower on the prestige scale. The chair of theoretical physics at Zurich remained unfilled all during the war. The department managed to function owing to the devoted services of its *Dozents*, Paul Epstein, Simon Ratnowsky, and Franz Tank. This arrangement saved the government some money, but in November 1919, Professor Meyer received permission to advertise an associate professorship in theoretical physics, and a selection committee was appointed. They decided it would not be possible to attract an eminent outside candidate and first considered the three *Dozents*.

By the end of January 1921, no decision had been reached, and meanwhile the committee had consulted Sommerfeld for his recommendations concerning a list of names, including Schrödinger. Sommerfeld replied in a detailed handwritten letter: "Epstein has doubtless among all your candidates the greatest name. His Stark-effect work belongs among the really greatest achievements. His general understanding and his mathematical aptitude are also remarkable. Yet it seems that in recent years he has become less productive." About Schrödinger, he said only: "A first-rate head, very sound and critical. Full professor at Breslau, and thus certainly not available to you [as associate]."

On February 2, Laue wrote that he would like to come back to Zurich, but his family had lost its money in the post-war inflation, and since he would have to rely entirely on his salary, it would have to be an exceptionally good one.

On February 5, a letter arrived from Furtwängler in Vienna: "About Herr Schrödinger as a person and as a teacher, there is only the most favorable to report. That he enjoys the greatest esteem here, you can see from the fact that he was ranked *primo loco* for the post in theoretical physics here. Unfortunately the ministry was not sufficiently forthcoming as to salary and he did not accept."

On February 26, Dean Wehrli went to Stuttgart to attend a lecture by Schrödinger and he asked him confidentially if he would accept a professorship at Zurich if it were offered. The answer was positive, and was confirmed in a letter of March 2.

In fact I was appointed at the University of Breslau on January 15 to a scheduled associate professorship with the personal rights (not salary) of a full professor. I take up this post on April 1. I would, however, *absolutely unequivocally give preference* to the professorship mentioned by you at the University of Zurich, if I should be called to it.

On March 3, the Dean wrote to the Education Commission that the first choice of the faculty was Laue and giving the other names in alphabetical order. Evidently the Directorate could not find the funds needed to attract Laue, and they sent the short list back to the faculty with a request that an order of recommendations be provided. Always on the lookout for savings, they also asked whether Professors Debye and Scherrer at the E.T.H. might not also provide for theoretical physics at the University. The task of the faculty was made easier by the sudden resignation of Paul Epstein, who accepted a post with Lorentz at the University of Leiden. Evidently his Polish accent and "too foreign ways" would be less of a problem among the Dutch.

On June 13, the Dean sent a thirteen-page letter to the Commission, summarizing the records of all the candidates, and giving the faculty's order of preference as (1) Schrödinger (2) Brillouin and Lenz, *aequo loco*, and (3) Ewald. He strongly argued that there was no way in which the E.T.H. professors could also provide for the teaching and research in theoretical physics at the University. At its meeting of July 20, the *Regierungsrat* of the Canton approved the recommendation of the Faculty and the appointment of Erwin Schrödinger as full professor of theoretical physics. In its report the faculty had pointed out

the versatility of the works of Schrödinger in the fields of mechanics, optics, capillarity, electrical conductivity, magnetism, radioactivity, gravitation theory, and acoustics. The appointment of Schrödinger was also deemed desirable by the Faculty because it would enable the holding of the lectures on biometry long desired by the biologists.

He was thus represented as a man conversant with all fields, but as yet he had not accomplished a truly outstanding piece of work in any particular field. To some extent the appointment was an act of faith that such wide-ranging interests and abilities would not go unrewarded by some major discovery. The duties of the post were specified as eight to twelve hours weekly of lectures and exercises in the field of theoretical physics, including a course of four hours a week on the mechanics of solid bodies every winter semester.

The annual salary would be 14,000 Sfr plus 30 percent of any tuition fees paid for his lectures and courses, after subtraction of costs. It is evident that the careful Swiss authorities left nothing to chance. The close control they exercised over the university administration at that time is noteworthy. The salary offered to Schrödinger, equivalent to about $2,500 a year, was at the top of the range for full professors. The country was in the midst of a post-war depression with severe unemployment in Zurich. The cost of living was fairly low since inflation in Switzerland had been much less than elsewhere. The cost of food had increased by 60 percent during the war years, with the best veal, for example, now at four francs ($.80) a kilo.

On September 16, Schrödinger wrote from Vienna to accept the appointment formally. At the age of thirty-four, he had achieved the first ambition of every European academic, a full professorship at an excellent university. As the successor of Einstein, Debye, and Laue, he must have felt inspired by a firm resolve to accomplish some physical research of the high standard set by these illustrious predecessors.

CHAPTER 5

Zurich

The city of Zurich is divided by the River Limmat into the newer *Grosse Stadt* on the left bank and the *Kleine Stadt* on the right bank, where the E.T.H. and University are located on heights overlooking the city, the river, and its source in the Zürcher See. The Schrödingers rented a spacious flat consisting of an entire floor of a modern stucco house called *Zu Vier Wachten* (At The Four Guards) at 9 Huttenstrasse, a broad avenue just a few blocks behind the University.

LIEGEKUR

Even before his lectures for the winter term began, Erwin was physically and psychologically exhausted. As he wrote about a year later to Wolfgang Pauli, "I was actually so *kaputt* that I could no longer get any sensible ideas. Not the least to blame for it were the many complications, the constant decisions about one's own fate, negotiations with ministries, etc. which I was not at all cut out for. Now that's all finished for a long time." He had scarcely started his lectures when he was forced by a severe attack of bronchitis to interrupt them in the middle of November. He suffered intermittent respiratory illnesses during the winter, from which he never managed to recover completely; finally a mild case of pulmonary tuberculosis was suspected, probably a recurrence of the infection that was noted two years previously in Vienna. He was ordered to undertake a complete rest cure (*Liegekur*) at high altitude.

The only bright aspect of this miserable winter was that his financial worries were over. Anny was able to engage a cook from Vienna, who expertly provided all their favorite dishes. This cuisine was a happy change from the near starvation of the post-war years in Germany and Austria.

The place chosen for Erwin's *Liegekur* was Arosa, an Alpine resort at about 1,700 meters altitude, overlooked by the great peak of the Weisshorn. Anny worried about Erwin like a mother with a sick child; she could not do enough to take care of him. Erwin improved remarkably during his nine-month stay in Arosa, and all evidence of active infection disappeared.

His return to Zurich was postponed till November 1, two weeks after the beginning of winter term; by then he had been pronounced cured, but he still tired easily. He resumed his heavy load of teaching but had little energy left for research. He had, however, managed to write two short papers at Arosa, of which the second was a truly original breakthrough into a new area, entitled "On a Remarkable Property of the Quantized Orbits of a Single Electron."

A REMARKABLE DISCOVERY

In the summer term of 1917, Hermann Weyl had given a course of lectures at the E.T.H., which were published the following year as *Space–Time–Matter*. The book gave an account of relativity theory including all the necessary mathematical background. It was very popular among physicists, and by 1921 was already in its fourth edition. Schrödinger studied this book assiduously and often used it in his subsequent work.

Weyl had shown that for a manifold to be a metric space it must have a measure of length at every point, and also every point must be metrically related to the domain surrounding it. This metrical relation requires that the length of a vector at the point P be related to its length after a congruent displacement to an infinitely near neighboring point. The immediate vicinity of P may be calibrated so that the length of the vector undergoes no change in such a displacement. Such a calibration is called *geodetic* at P. Schrödinger applied this measure theory to the orbits of electrons in the Bohr–Sommerfeld atomic models.

He considered the path of an electron in one completed orbit and showed that the Weyl condition for a geodetic path implied the existence of quantized orbits. Later he would realize that it contained the basis of de Broglie's explanation of Bohr orbits in terms of electron waves.

In 1926, Fritz London wrote a remarkably playful letter to Schrödinger about this paper:

Very Respected Herr Professor:
Today I must talk with you seriously. Do you know a certain Herr Schrödinger who described, in the year 1922, a "noteworthy property of quantum orbits"? Do you know this man? What, you say you know him rather well, you were even with him when he wrote this paper and were implicated in the work? This is truly shocking. Hence you already knew four years ago that one does not possess rods and clocks for the definition of an Einstein–Riemannian measure in the continuous description that occurs in analyzing atomic processes; thus one must see whether perhaps the general principles of measurement that arise from Weyl's theory of distance transfer might help. And you even realized four years ago that they help very well . . . and you showed that for real discrete orbits the gauge factor reproduces itself on a spatially closed path; and especially you then realized that on the nth orbit the unit of measure swells and shrinks n times, exactly as in the case of a standing wave describing the position of charge. You therefore demonstrated that Weyl's theory becomes reasonable . . . only if combined with quantum theory . . . You knew this and said nothing about it . . . Will you now immediately confess that, like a priest, you held the truth in your hands and kept it a secret . . .

The four years from 1922 to 1926 spanned a revolution in theoretical physics, so that what seemed clear to London in 1926 was only a clouded intuition to Schrödinger as he was fighting the bacillus at Arosa in 1922. Perhaps – had his health been better – he might then have realized all the implications of that strange property of electron orbits, and he would have discovered the wave properties of electrons before Louis de Broglie. Three years later, in this same mountain retreat, he would make his great discovery of wave mechanics.

LIFE IN ZURICH

Schrödinger had a heavy schedule of lectures for the academic year 1922–23, eleven hours a week. Unlike the situation today, the distinguished professors carried most of the teaching load.

Alexander Muralt recalled Schrödinger's classes during 1922–23:

At the beginning he stated the subject, and then gave a review of how one had to approach it, and then he started exposing the basis in mathematical terms and developed it in front of our eyes. Sometimes he would stop and with a shy smile confess that he had missed a bifurcation in his mathematical development, turn back to the critical point and start all over again. This was fascinating to watch and we all learned a great deal by following his calculations, which he developed without ever looking at his own notes, except at the end, when he compared his work on the blackboard with his notes and said "this is correct!" In summer time, when it was warm enough, we went to the bathing beach on the Lake of Zurich, sat with our own notes in the grass and watched this lean man in bathing trunks writing his calculations on an improvised blackboard which we had brought along.

Besides his colleagues at the University, Schrödinger developed close relations with the mathematicians and physicists at the nearby E.T.H. Pieter Debye, who had returned to Zurich in 1920, was about three years older than Erwin. Of all the theoreticians, he had the greatest insight into experimental problems and the best intuitive feeling about what was physically reasonable. He often met Schrödinger at the joint E.T.H.–University seminars; they were always cordial but never became close friends – solid Dutch virtues and fugitive Austrian charms made a somewhat immiscible combination. Erwin's closest friend when he first came to Zurich was Hermann Weyl, always called Peter by his intimates. Weyl, who was born in 1885 in Elmshorn, Germany, received the doctorate under Hilbert in Göttingen, and soon displayed mathematical genius of the highest quality. His wife, Helene Joseph, from a Jewish background, was a philosopher and literateuse. Her friends called her Hella, and a certain daring and insouciance made her the unquestioned leader of the social set comprising the scientists and their wives. Anny was almost an exact opposite to the stylish and intellectual Hella, but perhaps for that reason Peter found her interesting and before long she was in love with him. They were not a snobbish group

– some were rich and some were poor, and their favorite entertainments were not necessarily expensive.

Zurich is one of the oldest cities in Europe, the Roman town of Turicum having been established in 58 BC. It became famous during the Reformation under Ulrich Zwingli, a precursor of Calvin, and calvinism allied with capitalism remained the strongest influence in the life of the city. The Zurich *Frauenverein* was a league of right-minded ladies who tried to ward off any threats to the morals of their fellow citizens. During the war years, however, an influx of artistic and literary people formed an avant-garde minority culture, almost an underworld. Its symbolic representation was the *Steppenwolf*, created by Hermann Hesse, who when in Zurich lived with two dwarves in a house on the bank of the Limmat. A favorite meeting place was the Café Odeon where every table was the territory of some special group: ballet dancers, poets, anarchists, and so on. After the end of the war, as Europe counted its dead and the Swiss banks counted their profits, it was impossible to return Zurich to its pre-war calvinistic austerity. The *Frauenverein* had erected a wooden fence at the *Strandbad* to separate male and female bathers – one night it was torn down by rebellious students and never replaced.

Schrödinger liked to spend a leisurely hour or two sipping wine with friends at the "Bauschänzli," a tavern located on a little island in the Limmat, just before the Quaibrücke where the river joins the lake. When he wished for a more adventurous outing, he could rely upon Debye to provide an introduction to the less calvinistic circles of Zurich society.

Even in summer, the mountain snows never seemed far from Zurich. In winter, when cold fogs swept through the city streets, there was often bright sunlight above the clouds, and one could ascend nearby heights like the Uetliberg and escape from the urban gloom. When it was sunny on the heights, Zurich was filled with brilliant placards announcing "*Uetliberg Hell.*" Visitors sometimes thought they were advertising a light beer. Sunrise from the summit was a sight never to be forgotten, as the distant peaks appeared suddenly one by one, Jungfrau, Eiger, Monch, and Finsterhorn.

The year 1923 was a low point in Schrödinger's scientific production, for he published no papers at all. He had not completely recovered his health, and Debye recalls that he was not sick, not really sick, but very sensitive . . . He had quite a nice wife, but they never had children, for instance, which is also a little bit out of the order . . . He was rather nervous, but not nervous of the type of Zermelo. On the other hand he was quite more like a nice Austrian.

INAUGURAL LECTURE

On December 9, 1922, Schrödinger gave his inaugural lecture at the University of Zurich. Such lectures were intended for a public audience and hence supposed to deal with com- paratively general themes, often designed to show the philo- sophic view of the new professor towards his subject. While he was preparing this lecture, he began to worry again about the paradoxes of the wave-particle duality of light. He wrote to Wolfgang Pauli, who was then with Bohr in Copenhagen: "I for my part believe, *horribile dictu*, that the energy–momentum law is violated in the process of radiation." If light is emitted from an atom as a spherical wave, "Is there still a recoil there? Why? Can the conservation of energy–momentum not be merely a macroscopically valid average relation, of which atomic physics knows nothing, like the Second Law. At least it can be that way, and I see almost no other way out."

Schrödinger chose as the title of his lecture "What is a Natural Law?" harking back to a similar occasion in the autumn of 1907, when Franz Exner gave his inaugural address as Rektor Magnificus of Vienna University. Exner's thoughts had been inspired by the work of the late Ludwig Boltzmann, in which the laws of thermodynamics were derived from the statistical behavior of a myriad of interacting molecules. In 1919, Exner suggested that the individual molecular events which comprise a statistical ensemble may not be themselves subject to determinate laws, but may instead be completely random and devoid of any causal explanation.

This was the idea that Schrödinger expounded with enthusiasm in his inaugural lecture. It must have been an unusual notion at that time, since the principle of causality was then accepted by almost everyone. In the words of Cassirer: "Every natural event is absolutely and quantitatively determined at the very least by the totality of circumstances or physical conditions at its occurrence." Following Exner, Schrödinger denied that such determinism is necessary at the molecular level. For example, to maintain conservation of energy for a gas, it is not necessary that energy be conserved in each individual molecular collision; a sufficient condition is that in the average over many such collisions equal amounts of energy are gained and lost.

Even allowing for the fact that he was talking to a general audience, Schrödinger's treatment of the conservation laws was surprisingly superficial. As early as 1904, Georg Hamel had shown the relation of these laws to the fundamental symmetries of space and time. In mechanics, the principle of conservation of energy can be derived from the invariance of physical processes under a displacement in time. This invariance is such a fundamental part of physical reasoning that one is almost forced to infer that neither Exner nor Schrödinger was aware of the deep roots of the energy principle when they offered to eradicate it in such a facile fashion. Invariance with respect to temporal and spatial displacements is also the basis of Einstein's work on relativity, and conservation of energy and angular momentum is essential in the Bohr theory of atomic structure. Indeed Schrödinger admitted as much in his lecture: "I will not deny, however, that it is precisely the Einstein theory that in no uncertain terms makes plain the absolute validity of the energy–momentum principles."

His inaugural lecture is evidence of Schrödinger's enormous respect for the Vienna tradition of physics, and his desire to bring it to the attention of his new academic colleagues in Zurich. Nevertheless it is surprising that he devoted his first public appearance to the work of his old professor and not to any new or original ideas of his own. In the light of subsequent scientific history, these debates about causality and determinism in the early twenties may appear to foreshadow the much more cogent attack on causality derived from quantum mechanics by Werner Heisenberg in his indeterminacy principle of 1927.

Schrödinger had found in Hermann (Peter) Weyl a kindred spirit in many ways, including their mutual distrust of causality. As a *Dozent* at Göttingen, Weyl had been influenced by Edmund Husserl (1859–1938) who held the chair of philosophy there from 1901 to 1916, and his wife Helene Fischer had been a student of Husserl. Husserl was the creator of a philosophy called "phenomenology" (not to be confused with "phenomenalism") in which psychological introspection largely replaces any empirical study of an external world; all things that appear in consciousness are worthy of attention and analysis, whether or not they are metaphysically real, and perhaps the phenomenological structures of the mind are even more interesting when they happen to be unreal. In 1920, Weyl published a paper "On the Relation of the Causal to the Statistical Consideration of Physics," in which he rejected causality because of its incompatibility with our intuition of the inwardness of time. He would have been sympathetic also with Schrödinger's interest in eastern religions, and especially with the concepts of tantrism, in which all creative activities in the cosmos are allied to the creativity of human erotic experiences.

Schrödinger may have had second thoughts about his inaugural address, since he did not allow it to be published until six years later, by which time he had become a famous scientist who could afford to express radical ideas. Also by that time he had changed his views drastically and had become an outspoken opponent of the concept that physical events at the atomic level are random manifestations of pure chance.

ZURICH 1923–24

Hunger and starvation were still widespread in Germany. The allied blockade had been maintained after the Armistice, and thousands perished in the cities. In Switzerland, collections and benefits were organized for German Relief. Zurich citizens contributed to maintain a soup kitchen in Stuttgart that fed thousands of people every day. At this time one of the main stories in the *Neue Zürcher Zeitung* was the drawn-out trial in Munich of Hitler and Ludendorff as ringleaders of the abortive beer-hall putsch.

Schrödinger was happy with the natural environment and the intellectual life of Zurich, but he was growing older and still had not made any major contribution to physics. As a schoolboy and university student, he had always been at the top of his class, but now he had to watch younger contemporaries forging ahead of him. He was well aware of sentiments like that expressed by Dirac in a student ditty:

> Age is of course a fever chill
> That every physicist must fear.
> He's better dead than living still
> When once he's past his thirtieth year.

Except for *Farbenlehre*, which was outside the mainstream of physical theory, there was no division of physics in which Schrödinger was an acknowledged leader. When he ventured to propose an original theory, the model usually turned out to be naive and untenable. If Schrödinger had died in 1924 at the age of thirty-seven, his work would have merited only a footnote in the history of modern physics. In 1924, however, he regained his research momentum and published six papers, although none were of major significance.

Erwin's personal life was entering a period of turbulence. He and Anny had become members of a close group of friends, almost a clique, that included the Weyls, the Bärs, and the Meyers. The families spent many hours together in picnics and swimming parties. Erwin became popular among the children for his ability to stand on his head, but he was unhappy that he had no offspring of his own – he had always wanted a son so that he could re-enact his own close relation with his father. Although he had enjoyed several casual love affairs, they had not been serious, and he was becoming unhappy with Anny's evident interest in other men. They began to consider the possibility of a divorce. Thus, as the year 1925 began, and his physical health approached normality, his psychic state was far from tranquil. It would prove to be a marvelous, almost a miraculous year for his theoretical physics.

THE NATURE OF LIGHT

As a result of contacts and discussions with European colleagues, Schrödinger was by now deeply concerned with problems in atomic physics and quantum theory, especially the nature of radiation and how it interacts with electrons and atoms. These new interests brought him into closer relations with work in progress in the schools of Sommerfeld in Munich, Born in Göttingen, and especially Bohr in Copenhagen, all of which had close connections with one another.

In 1916, Bohr had returned to the University of Copenhagen to a chair of theoretical physics that had been especially created for him. Although he was still only thirty-one years old, his subsequent influence was exerted not so much through original research as through his inspiration of others and his provision for them of an ideal climate for intellectual work. He also began to delve into the logical and philosophical foundations of physics. Even his first papers from Copenhagen were often essays in search of verbal understanding, rather than mathematical analyses of crucial problems. His lecturing style was a discursive mumble, but with small groups and especially in one-to-one discussions he was without an equal in his enthusiasm, his empathy, and his contagious love of his subject. At the end of the war, young physicists from all over the world flocked to Copenhagen, where science could be pursued in an atmosphere unpoisoned by politics. In 1922, Bohr was awarded the Nobel prize for "his services in the investigation of the structure of atoms and the radiation emanating from them."

As early as 1912 it had been noted that when a beam of X-rays of well-defined wavelength is scattered by material of low atomic weight, some of the scattered X-rays have a longer wavelength than before. In 1922, Arthur Compton at Washington University (St. Louis), explained this effect by treating an electron and a light quantum (photon) as two particles and applying the laws of conservation of energy and momentum to a collision between them. This paper brought the particle theory and the wave theory of light to a decisive confrontation. Sommerfeld wrote to Compton that his work had sounded the death knell of the wave theory. Yet interference phenomena required waves of radiation. It seemed as if every physicist was now summoned to take a stand concerning the nature of light: either for waves or for particles.

John Slater was twenty-two years old and had just taken his Ph.D. at Harvard. Uninhibited by the factions in European physics, he asked himself, "Why must it be either – or? Why can't radiation be both wave and particle?" He took this heretical idea to Copenhagen at Christmas 1923. Slater proposed that all the time an atom is in an excited state, it is emitting electromagnetic waves of all the frequencies corresponding to transitions to lower states allowed by Bohr's theory. These electromagnetic waves, however, were of a peculiar kind; they did not carry energy, but were connected with the probability of finding photons at a given point.

Bohr and his assistant Hendrik Kramers were enthusiastic about this idea, but they proceeded to extend it in ways that made Slater uneasy. Bohr coined the term "virtual oscillations" for the electromagnetic waves. The dictionary defines "virtual" as "existing in essence or effect though not in actual fact." Slater was willing to accept this term, although he did not in actual fact understand it, but Bohr and Kramers went much further: they did not include the actual existence of light particles in the theory, and they abandoned the conservation of energy and momentum as applied to an individual emission or absorption of light, retaining only a statistical validity for these laws. They may have been encouraged in this folly by learning of Schrödinger's views from Pauli, but Einstein found this idea "to be disgusting."

The Bohr–Kramers–Slater paper tried to describe the Compton effect without photons but the attempt was not convincing. It predicted that the recoil electron and the scattered quantum would not appear simultaneously in Compton scattering. Experiments by Walther Bothe and Hans Geiger in

1925 demonstrated a simultaneous process, as did observations by Compton and Simon with a Wilson cloud chamber. Schrödinger's reaction to the B-K-S paper was enthusiastic. In a letter to Bohr, dated May 24, 1924, he stated that:

I have just read with the greatest interest the interesting change in your ideas . . . I am extremely sympathetic to this change. As a pupil of old Franz Exner, I have long been fond of the idea that the basis of our statistics is probably not microscopic "regularity," but perhaps "pure chance."

In the September 5 issue of *Die Naturwissenschaften* Schrödinger published a paper on "Bohr's New Radiation Hypothesis and the Energy Law." He was excited by

the fundamental violation of the energy–momentum laws in each radiation process. This violation is not something trivial . . . it seems unavoidable that the entire system experiences considerable, completely irregular changes in its energy content, like a gambler who in rapid succession bets a large part of his funds on a game of chance that he has a 50–50 chance of winning, but just this comparison solves the difficulty, since he can bet a lot each time in a huge number of small bets.

The Schrödinger paper was too much for Bohr, however it may have been intended; as he pondered its implications and the results of the Bothe–Geiger experiment, Bohr realized that the B-K-S paper could no longer be taken seriously. The paper was the last major effort to deal with the interaction of radiation and matter in terms of the old semi-classical quantum theory. Henceforward the theoretical physicists would seek more radical answers. They were ready for a scientific revolution.

INNSBRUCK

The eighty-eighth Meeting of German Scientists and Physicians was held in Innsbruck from September 21 to 27. Egon Schweidler, Erwin's former colleague in Vienna, was in charge of the arrangements. Arthur March, a *Dozent* at the University of Innsbruck, was a member of the local committee. It was a large meeting, with important lectures by Sommerfeld, Ewald,

Planck, Laue, Born, and others. Schrödinger was there but did not present a paper, nor did he contribute to the published commentaries, but he was an enthusiastic participant in the informal discussions and social events. He was happy to be back again in Austria and to renew old friendships with the Austrian physicists.

Early in 1925 the Professorship of Theoretical Physics at the University of Innsbruck became vacant. On June 6, the Philosophical Faculty recommended to the Austrian Ministry for Education that Schrödinger be appointed to the Chair. The nomination of Schrödinger included a reference to his experimental work, since it was desirable

that the post be filled by an investigator of many-sided abilities, who has an overview of the entirety of his science and therefore also a feeling for empirical research. Our university would not be a suitable base for one-sided specialists, who are devoted exclusively to purely mathematical formulation or – an increasingly common type at this time – who pursue speculations far removed from reality.

As proof of his versatility, the forty-two publications of Schrödinger were cited: mechanics (4), atomic physics (11), radiation theory (10), color theory (7), geophysics (2), and experimental investigations (4).

On July 21, 1925, Schrödinger wrote to Sommerfeld: "I nearly forgot to tell you – if you don't know about it – that I am expecting to receive a call to Innsbruck. I should like to return home, because the Swiss are *gar zu ungemütlich*." On January 15, 1926, an Innsbruck newspaper reported that "in times of great financial stringency, the Vienna ministry is preparing to select the most expensive teacher. It is about to call Professor Dr. von Schrödinger [sic] from Zurich and the ministry is prepared to find an exorbitant salary of 1,200 schillings a month to pay him." Another paper complained that "five state positions with an average pay of 300 Sch must be abolished in order to afford this costly luxury call to a small provincial university." On January 29, 1926, Schrödinger wrote again to Sommerfeld:

Officially I have not yet decided about Innsbruck. I do believe, however, that I shall stay here. It is mainly the fact that Schweidler is leaving for Vienna that decides the case for me . . . Please do not mention my decision as being certain when you talk to anybody. It would be unpleasant for me in dealing with both ministries. On the other hand, the delay is perfectly fine with me, because now I shall obtain here after much effort some little improvements, that is, a new blackboard in the lecture hall and, hopefully, a little larger budget for the seminar library.

Schrödinger officially declined the Innsbruck offer about March 16, and on March 25 he received a Sfr 1,000 increase in salary. By this time, the first of his great papers had appeared and he was about to be recognized as one of the world's most eminent theoretical physicists.

MICHELSON–MORLEY

In addition to all his other activities in 1925, Schrödinger became involved in a controversy arising from the degenerate state of German politics. Reactionary groups in Germany refused to accept the reality of the defeat of the Kaiser's military machine, and sought to exploit an historical fantasy in which an undefeated army was betrayed by a sinister alliance of Jews and socialists. The reactionary forces included remnants of the military, industrialists and large landowners, and many university professors. They all hated the social-democratic government of the Weimar Republic and were dedicated to its destruction, the generals and business magnates by overt actions and the university mandarins by incessant and insidious propaganda.

A particular focus of their hatred was Albert Einstein. His pacifist and socialist views were well known; he had refused to sign the statement of German professors supporting the invasion of Belgium, and had consistently advocated a negotiated peace even when Germany seemed to be winning the war. The German physicists were generally conservative but not extremist, except for a small faction led by Philipp Lenard and Johann Stark, both Nobel prize winners. Like Hitler, Lenard was an Austrian, and he was at first not rabidly anti-semitic, but merely in the typical Austrian way.

An important experimental support for Einstein's special theory of relativity was the null result of the Michelson–Morley experiment of 1887: the failure to detect any motion of the earth relative to a fixed aether. In 1921, Dayton C. Miller, who had been a junior colleague of Michelson at the Case Institute in Cleveland, performed a Michelson–Morley type experiment at the top of Mount Wilson, and reported a positive result, a small but significant effect of the earth's motion on the velocity of light. He theorized that at sea level the aether is carried along by the earth, whereas at higher altitudes a small aether wind becomes evident. Einstein was visiting Princeton when he heard of these results and made his statement, "Raffiniert ist der Herr Gott, aber boshaft ist er nicht." (The Lord God is subtle, but he is not malicious.) The Miller result was received with delight by the enemies of the Einstein theory but, according to Schrödinger, surprisingly Lenard did not believe it.

On April 1, 1924, Adolf Hitler was sentenced to prison for his part in the Munich beer-hall putsch, and Lenard and Stark published a statement in his support: "One ponders how much it means that we are allowed to have this type of spirit living amongst us in the flesh . . . such spirits are embodied only with Aryan-Germanic blood." Henceforth, Lenard's Heidelberg institute became a center for Nazi agitation. Such was the background of the plans for a co-operative effort of some German and Swiss scientists to arrange a replication of the Miller experiment at a much higher altitude. Schrödinger wrote to Willy Wien, a moderate anti-semite, on September 17, 1925: "The result of the Miller experiment is very important, but it has been played down in Jewish circles of physicists. I should like to see the experiment repeated on the Jungfraujoch." (3,457 meters altitude). Wien found that Rudolf Tomaschek who had worked with Lenard at Heidelberg would be willing to do the experiment; he was highly qualified, having already published excellent interferometric studies.

At this time, Schrödinger's experimental colleague at Zurich, Edgar Meyer, intervened, saying that he would not trust any of Lenard's people to be objective about the work. Erwin was incensed, and reported to Wien on December 27:

Please forgive me that I have not answered you before on the Tomaschek matter. I told Edgar Meyer that you had informed me that Tomaschek was prepared to do the experiment. The astonishing answer of Meyer was that he knew this long ago. And now came a bunch of "buts and howevers" which finally came down to the point that Tomaschek as a pupil of Lenard on account of the antagonism against Einstein . . . would not be suitable. Besides Joos may be obtained and that would be better yet. I don't like to complain about a colleague, but the whole way in which this affair has been managed doesn't please me at all.

In the event, Georg Joos carried out an excellent series of experiments on the Jungfraujoch, financed jointly by the Germans and the Swiss. The results confirmed the absence of any detectable "aether wind," and Einstein's theory thus received further experimental support.

In this controversy, Schrödinger aligned himself with the more nationalist and anti-semitic wing of the German physics community. Actually, he was basically contemptuous of politics, and he was correct in his view that Tomaschek was a trustworthy scientist, who should not be held responsible for the political activities of his professor.

MY WORLD VIEW (*MEINE WELTANSICHT*)

In the autumn of 1925, Erwin Schrödinger wrote an intensely personal account of his philosophy of life. He did not publish it at that time, perhaps he was just then swept up in the onrush of creativity that led to wave mechanics, but perhaps also because this book revealed his innermost heart in a way that might seem immodest for a physicist yet to gain a place among the immortals. The testament of 1925 must have been the result of many days of meditation: here am I, thirty-eight years old, well past the age at which most great theoreticians have made their major discoveries, holder of the chair that Einstein once held, who am I, and what do I believe?

The 1925 part of *Meine Weltansicht*, called "In Search of the

Way," consists of ten short chapters. In 1960, the year before his death, he wrote five more chapters, and published them all together. The earlier chapters, which are considered here, state his view of the desperate state of metaphysics, provide a means of rescue based upon the philosophy of Vedanta, and conclude with reflections on several psychobiological themes such as inherited memory.

Schrödinger begins with a consideration of "Metaphysics in General": since Kant, it is easy enough to blow away the structure of metaphysics like a house of cards, but having done so, one is left with the empty feeling that art and science are no longer worthy of serious dedication. The task of post-Kantian scientific philosophy is to understand the contradiction between the necessity of a metaphysical framework for scientific endeavor, and the flimsy unscientific structure of the framework itself.

The "death of metaphysics" has led to an overdevelopment of technology and a decline of the arts. The churches, guardians of the most holy treasures of mankind, have wasted their spiritual resources, and "the sparks of ancient Indian wisdom, kindled to a new glow by the wonderful Rabbi by the Jordan, have been almost extinguished . . . The middle classes have become unprincipled and leaderless . . . A general atavism has set in . . . crass, unrestricted egoism lifts its smirking head and with its primeval irresistible iron fist seizes the rudder of the leaderless ship." This chapter reveals a man profoundly disturbed by the decline of western civilization – and the worst by far was yet to come. It is a cry of spiritual pain of a soul torn between a need for religious belief and an inability to accept such belief without treason to his intellectual standards.

Schrödinger wants to make a statement about the basic concepts of philosophy, but he feels held back because he knows the lack of clarity, the falsity, the one-sidedness of every such statement. Thus the situation calls for the Buddhist wisdom that sought to express the unexpressible through contradictory statements: a thing can be either A or not-A, but can also be both A and not-A at the same time.

Chapter 5, "The Basic View of Vedanta," is the culmination

of the book. Here Schrödinger attempts to find a solution to the problem raised by the multiplicity of observing and thinking individuals. Vedanta teaches that consciousness is singular, all happenings are played out in one universal consciousness, and there is no multiplicity of selves. Schrödinger does not believe that it will be possible to demonstrate this unity of consciousness by logical argument – one must make an imaginative leap, guided by communion with nature and the persuasion of analogies. The epigraph to this statement of his beliefs is from Goethe:

> Und deines Geistes höchstes Feuerflug
> Hat schon am Gleichnis, hat am Bild genug.

And thy spirit's highest fiery flight / Is satisfied with likeness and with image.

A brief summary of Vedanta was given by George Thibaut who translated the Sutras with the commentary of Śankara in the great series, *The Sacred Books of the East*, edited by Max Muller. The Sutras are brief, often cryptic statements, which require interpretation to become intelligible. The interpreters may differ among themselves on important points, but Śankara provides a standard and orthodox view of Vedanta. In his opinion, the Upanishads teach the following:

There exists only one universal being, called Brahman, which comprises all of reality in an undivided unity. This being is absolutely homogeneous in nature. It is pure thought. Thought is not an attribute of Brahman, but constitutes its substance. Thus Brahman is not to be called a thinking Being, rather it is Thought itself. It is absolutely destitute of qualities.

Whence then comes the phenomenal world of ordinary experience, and our own individual selves and qualities? The answer is that Brahman is associated with a certain power called Maya to which is due the appearance of the entire world. Maya is neither being nor not-being, but a principle of illusion. As a magician creates illusions on a stage, Brahman through Maya projects the appearances of the world. Maya is the material cause of the world. In all the apparently individual forms of existence, the indivisible Brahman is present. All that is real in any individual soul is Brahman, the appearances are unreal. External things are Maya, the thoughts we have of them are Maya, neither one is more or less real the other. Thus Vedanta is not an idealist philosophy in a technical sense. The unenlightened soul is not able to look through or beyond Maya. It is engaged in an endless *samsara*, a cycle of birth, action, and death. The soul, which in reality is infinite Brahman, pure thought, is enmeshed in the unreal world of Maya. The only way out is provided by the Veda. Devout meditation on the Sutras can finally lead the soul to know that there is no difference between its true self and the highest Self. It will then be released from Maya and will realize its true identity with Brahman. Schrödinger says: "This is what Brahman expresses with the holy, mystical, and yet actually so clear and simple formula '*Tat Twam Asi* – This Thou Art' or also with such words as 'I am in the East and in the West, I am below and I am above, *I am this entire world*'." This is not meant in Spinoza's pantheistic sense, that you are a part, a piece, of an eternal infinite being, since then the old question would arise, what piece are you? "No, as inconceivable as it seems to ordinary reason: you – and every other conscious being taken in itself – you are all in all . . . it is a vision of this truth that forms the basis of every morally valuable activity."

Such a vision of the unity of the world – nature, man, and God – has not been confined to eastern seers. From India it spread to the neo-platonic Greek philosophers, notably Plotinus, for whom the universe is a living organic whole, which emanates from the One or the Good, the divine Intellect. In the words of A. H. Armstrong: "In this divine Intellect, thought and its content are one, so that it can be regarded either as a unity in a diversity of forms, or as a unity in a diversity of minds, each of which thinks and so is the whole. In terms of our consciousness, Intellect is the level of intuitive thought that is identical with its object and does not see it as in some sense external."

The ancient Hindu religion was reborn in the West in various forms of Gnosticism. The Sanskrit word "Veda" and the Greek "Gnosis" both mean "Knowledge." The gnostic doctrine is that salvation is to be achieved through secret knowledge. While the apostles were spreading the gospel of Christ, Simon the Magician was wandering through Asia Minor with his faithful companion Helena, whom he had redeemed from a brothel in Tyre. He preached an elaborate gnostic mythology, in which the harlot typified the female holy spirit of wisdom, Sophia or Ennoia. Her representation as a whore showed the depth to which the divine had fallen by becoming involved in the creation of a material world.

The secret knowledge of gnosticism is knowledge of the self. Schrödinger had read the Christian mystics, who found God in the emptiness of their innermost souls. A favorite metaphor was "the inner eye." Thus Jacob Böhme said: "The unity is an eye which looks, which sees itself, which creates vision," and Johann Eckhart told us: "The eye with which Man sees God is the same as the eye with which God sees Man." In modern gnosticism, "The universe is like the visual field of a living brain."

Vedanta and gnosticism are beliefs likely to appeal to a mathematical physicist, a brilliant only child, tempted on occasion by intellectual pride. Such factors may help to explain why Schrödinger became a believer in Vedanta, but they do not detract from the importance of his belief as a foundation for his life and work. It would be simplistic to suggest that there is a direct causal link between his religious beliefs and his discoveries in theoretical physics, yet the unity and continuity of Vedanta are reflected in the unity and continuity of wave mechanics. In 1925, the world view of physics was a model of the universe as a great machine composed of separable interacting material particles. During the next few years, Schrödinger and Heisenberg and their followers created a universe based on superimposed inseparable waves of probability amplitudes. This new view would be entirely consistent with the vedantic concept of the All in One.

The remainder of Schrödinger's 1925 *Meine Weltansicht* reflects his interests in evolutionary biology and mechanisms of heredity. It relies heavily on the work of Richard Semon, and thus harks back to his early days at Vienna University, and his long discussions with his friend Fränzel, the religious student of biology. Almost twenty years have passed since then, but maturity and experience have not diluted his enthusiasm for Semon and his *mneme*.

At the time this book was written, Erwin's marriage was close to disintegration. After four years together, he and Anny had no children, and their sexual incompatibility was such that it was unlikely that they ever would. The special circle in which they lived in Zurich had enjoyed the sexual revolution a generation before Alfred Kinsey discovered it in Indiana. Extramarital affairs were not only condoned, they were expected, and they seemed to occasion little jealous anxiety. Anny would find in Hermann Weyl a lover to whom she was devoted body and soul, while Weyl's wife Hella was infatuated with Paul Scherrer, Debye's colleague at the E.T.H. Erwin and Anny considered the possibility of a divorce, but the idea did not appeal to Anny. Outside the bedroom, they were a compatible couple; she was a competent housekeeper and a faithful correspondent. Thus they agreed to continue the marriage, while each remained free to find sexual relationships elsewhere. In view of Erwin's desire for a son, this was a curious arrangement, but Anny was nominally a Catholic and divorce would have left her in a precarious position. Furthermore, divorce was expensive, and this factor must have had some weight, given Erwin's financial prudence. The principal factor, however, was the romantic tendency of Erwin's character. He was convinced that bourgeois marriage, while essential for a comfortable life, is incompatible with romantic love. On the evidence of his diaries, he was not a libertine for whom sexual conquest was an aim in itself – it was the falling in love that he valued most. Several of his greatest loves never led to sexual union, but the romantic passion was valued for its own sake.

This personal background may explain a certain aridity that pervades Schrödinger's "World View." There is little joy in it and no love at all. An abstract contemplation of the continuity of inherited behavior is a bleak substitute for the happiness of shared family life. The unity of consciousness in Brahman, especially as a vicarious experience, provides no guidance through the actual conflicts of egotistical personalities. Even if we share our minds with all other mortals, with our dogs and cats, and even with flowers and crystals, there is limited consolation in such theoretical communion. Erwin, however, never allowed his distrust of Maya to inhibit his enjoyment of its illusory pleasures. He found happiness in his intellectual work, the companionship of friends, good wine, poetry and drama, the love of women, and explorations of mountains and seashores. In these ways also he resembled his mentor Schopenhauer, who was able to combine a pessimism about the world with an indulgence in its pleasures, and an almost paranoid misogyny with an ardent pursuit of fair women.

The philosophy of Schrödinger at this time does not appear to have been influenced by his physics. In fact he often said that one cannot derive philosophical conclusions from physics. In contrast, however, he was willing to admit that philosophy could influence physics. Like Schopenhauer, he accepted an hierarchical view of our understanding of the world, with philosophy above and physics below. The philosopher once said that "physics is unable to stand on its own feet, but needs a metaphysics on which to support itself, whatever fine airs it may assume towards the latter." Perhaps this is why, as a young man, Erwin considered giving up serious physics to become a philosopher. Yet, in *Meine Weltansicht* he does not attempt to provide a metaphysical foundation explicitly designed for physics.

At this time, his philosophy of science (as distinct from his philosophy in general) was derived in part from Mach and in part from Boltzmann, and any synthesis of these conflicting elements must have seemed impossible. Moreover, both physics and its philosophy were on the verge of revolutionary changes. The influence of Vedanta on Schrödinger's physics is obscured by the fact that he was a reluctant revolutionary. Thus others would argue that quantum mechanics brought the world view of physics closer to that of Vedanta, but Schrödinger, the most consistent follower of Vedanta among physicists, refused to derive this consequence from his own discoveries.

GAS STATISTICS

Besides his work on color theory, Schrödinger's principal research during his early years in Zurich was on the statistical thermodynamics of ideal gases. This field was part of the great legacy of Boltzmann, which related the macroscopic to the microscopic world. Schrödinger's choice of a statistical theme for his inaugural lecture showed his abiding interest in such problems.

Thermodynamics is based on three fundamental postulates or Laws. The First Law is the principle of conservation energy: energy can be neither created nor destroyed in any process – the energy of the universe is constant. This law, however, tells nothing about the direction of changes in the world, yet natural processes occur in particular directions, for instance, heat flows from hotter to colder bodies, gases mix by inter-diffusion. The energy function cannot predict the direction of a change, since it always remains constant in isolated systems or in the universe as a whole. We need a function that changes when the system changes and stays constant when the system rests at equilibrium. Midway through the nineteenth century, Rudolf Clausius in Zurich and William Thompson in Glasgow postulated the Second Law of Thermodynamics, and Clausius used it to define a new function, which he called *entropy*, from the Greek for "in change." In all the large-scale changes that occur in the universe, the entropy always increases.

Why does the entropy of the universe always increase in any naturally occurring process? The greatest achievement of Ludwig Boltzmann was to give a plausible answer to this question in terms of the motions of the enormous numbers of particles that make up any large-scale material system. He saw that the increase in entropy of a system means that the particles that compose it are moving from less probable to more prob-able arrangements. The equilibrium state, in which the entropy has reached a maximum, corresponds to the state of maximum probability. Boltzmann set the probability propor-tional to the number of different "complexions," W, of the particles in a system, for example, the number of ways in which a total energy E can be distributed among N molecules. Following Boltzmann's reasoning, Max Planck obtained the famous relation

$$S = k \log W$$

where k is the Boltzmann constant and $\log W$ the natural logarithm of W. In applications of this equation the quantum theory was to play an important role, since it could provide the essential information on the allowed energy states of various systems.

Another entropy problem of great interest was found in the experimental studies of Walther Nernst in Berlin. In 1906, he proposed his "Heat Theorem," which was soon recognized as the Third Law of Thermodynamics. One statement is that in the limit of absolute zero $(0\,\mathrm{K})$, the entropy change

$$\Delta S = 0$$

for any process. Furthermore, the heat capacity C_v of any substance goes to zero in the limit of $0\,\mathrm{K}$. At first, Nernst restricted these theorems to liquids and solids, but later he saw that they are also valid for gases. Attention was directed to what was called "the degeneracy of gases" at very low temperatures, by which was meant the Third Law requirement that their heat capacities should approach zero.

Early in 1924, Schrödinger entered this field with the publication of a paper on "Gas Degeneracy and Free Path Lengths." As had occurred so often in the past, his model lacked physical realism and he abandoned it a few months later. His interest in the counting problems swirling around the

$$S = k \log W$$

formula was, however, thoroughly aroused. His next attack on the entropy problem was a non-mathematical paper, an exercise in pure logical analysis: "Remarks on the Statistical Definition of Entropy for an Ideal Gas." He sent this to Max Planck for submission to the *Sitzungsberichte* of the Prussian Academy of Sciences, where it was reported in July 1925.

In the eighteen months since his previous paper, there had

been a major advance in the statistical theory of gases. In June 1924, Satyendra Nath Bose, a young Bengali physicist at the University of Dacca, sent Einstein a letter with a copy of a paper that had just been rejected by an English physics journal. The paper contained a new derivation of Planck's radiation law. Einstein was so impressed with the paper that he translated it personally and sent it to the *Zeitschrift für Physik*, with a note saying "In my opinion, Bose's derivation of the Planck formula constitutes an important advance. The method used here also yields the quantum theory of the ideal gas, as I shall discuss elsewhere in more detail." Bose derived the Planck distribution law for thermal radiation without any reference to electromagnetic theory, but simply by applying an appropriate statistics to the photons in a container of volume V. He considered the photons to be massless particles capable of existing in two states of polarization, with momenta given by

$$p = h\nu/c.$$

They were assumed to be indistinguishable and their number N was not necessarily conserved. All three of these assumptions are correct.

Einstein saw immediately that the molecules in a gas are also indistinguishable particles to which the Bose method of counting can be applied. The difference from photons is that the number of particles N must be held constant in calculating the maximum of W. Einstein published three papers on the subject, the first in September 1924, and the last two early in 1925. In his second paper, he wrote "If it is justified to conceive of radiation as a quantum gas, then the analogy between the quantum gas and the molecular gas must be a complete one." Einstein believed that the relation was in fact more than a mere analogy, and that molecules as well as photons must have both particle and wave characters. He cited the work of Louis de Broglie at this point, thus making a connection between quantum statistics and the wave properties of matter that would certainly have attracted the attention of Schrödinger during his reading of the Einstein papers.

Although the Bose–Einstein statistics gives a correct general counting for the microstates of indistinguishable molecules,

Schrödinger showed that when temperature is not too low and density not too high, the number of states is so much greater than the number of molecules that N_i, the number of molecules per state, almost always equals either 0 or 1. States with $N_i > 1$ are so rare as to be negligible. Under these conditions, division of the classical counting of W by $N!$ gives the same result as the quantum statistics. This paper of Schrödinger's was in a sense the last word on a difficult and controversial subject until quantum mechanics showed the necessity for two different kinds of statistics, depending on the symmetry of the particle wave functions.

On September 26, Einstein wrote to Schrödinger to say that he had enjoyed his Prussian Academy paper on entropy. He went on to suggest a way in which Planck's idea of treating the energy levels of the entire gas instead of those of individual molecules could be formulated. On this basis Schrödinger calculated explicit but impractical formulas for a sort of "semi-ideal" gas. He sent an outline of this paper to Einstein and asked him to be co-author of a joint publication. Einstein replied that he did not believe he should be a co-author since Schrödinger had done all the work, and he would not wish to appear as "an 'exploiter' as the socialists so prettily put it." Schrödinger then sent him the completed paper with the authors' names omitted. "Even in jest, I would not consider you as an 'exploiter'. To remain with the sociological framework, one could say instead: When kings build, the hodcarriers have something to do."

Einstein put Schrödinger's name on the paper and sent it to the Prussian Academy in February 1926. Einstein was wise not to put his name on this paper – it would have added nothing to his three great papers on the subject.

PARTICLES AND WAVES

The discovery that laid the foundation for Schrödinger's revolutionary papers in wave mechanics came from an unexpected source, a thesis presented by a young French physicist for a doctoral degree at the University of Paris.

Louis de Broglie was born on August 15, 1892, in Dieppe,

the youngest of five children. His father, a member of the French parliament, died in 1906 and his elder brother Maurice, who was seventeen years his senior, became head of the family and responsible for his education. They lived in a mansion in Paris, where Maurice had a private laboratory for research on X-rays. At first, however, Louis was more interested in history than in science, but in 1911 after Maurice told him about the exciting developments in quantum theory and relativity, Louis decided to study physics. In 1913, he received his *Licence ès sciences*, equivalent to a B.Sc. degree, and entered the army for his year of military service. When war broke out, he was assigned to the radiotelegraphy unit in the *Tour Eiffel*. He remained there for five and a half years, learning an enormous amount of practical electromagnetism.

It is interesting that both de Broglie and Schrödinger were working on the theory of ideal gases just prior to their respective discoveries of wave-particle duality and wave mechanics. De Broglie recalled that "all of a sudden" he saw that the so-called crisis of optics was simply due to a failure to understand the true universal duality of wave and particle. He first published the basic elements of his discovery in three short notes in the *Comptes rendus* of the Paris Academy in September and October, 1923, and a more elaborate version in his thesis for the doctorate of science, defended on November 25, 1924. He derived a relation between the wavelength λ of the particle and its momentum *mv*:

$$\lambda = h/mv.$$

A copy of the thesis was sent to Einstein who at once recognized its importance and wrote to Langevin: "*Er hat eine Ecke des grossen Schleiers gelüftet.* [He has lifted a corner of the great veil.]"

INTIMATIONS OF WAVE MECHANICS

In the summer semester of 1925 Schrödinger lectured on "Quantum Statistics." Many scientists find the preparation of a course of advanced lectures to be an inspiration for further

11. Louis Victor de Broglie (c. 1924)

research in the subject concerned, as they must read all the current literature and in explaining it to students often become aware of gaps in the theory or in their own understanding. Such was the background of his paper "On Einstein's Gas Theory," submitted on December 15. Since this was the last paper that he wrote before the explosion of creativity in his discovery of wave mechanics, it has attracted much attention as an intellectual precursor of that discovery. The paper demonstrates how seriously he was thinking about particles as waves. Schrödinger later said that "wave mechanics was born in statistics," and it may have been conceived during this work on the statistical mechanics of the ideal gas.

This paper is the best of his contributions to quantum statistics, but it still has the character of his earlier works, a critical reaction to an idea proposed by someone else, followed by an attempt to refine its mathematical analysis and to sharpen its theoretical relevance. In this case, he took a method of evaluating statistical probabilities, given in 1922 by Charles Darwin and Ralph Fowler, and used it for a new derivation of the Einstein–Bose gas statistics.

Louis de Broglie had emphasized that every particle has the wave properties of wavelength and frequency. In his paper on gas statistics, Einstein had accepted wave-particle duality, and many physicists first learned of this concept through him, and were obliged to take it seriously by the weight of his authority. Schrödinger must have studied the de Broglie papers carefully in the summer and fall of 1925. He says that his approach to gas statistics "means nothing other than taking seriously the de Broglie–Einstein wave theory of the moving particle, according to which the latter is nothing more than a kind of 'whitecap' [*Schaumkamm*] on the wave radiation that forms the basis of the world." This statement goes beyond anything that de Broglie ever wrote about the relation between waves and particles. It is the first expression by Schrödinger of what was to become a major theme in his interpretation of wave mechanics: the world is based on wave phenomena, while particles are mere epiphenomena.

Schrödinger is now taking the wave nature of matter very

seriously, beginning to think in detail about the kind of wave that is required and the laws and equations that it must obey. He is formulating the dispersion laws in relativistic terms, and thus it will be only natural that his first efforts to find a suitable wave equation will also be based on the relativistic equations. The final section of his paper considers "The Possibility of Representing Molecules or Light Quanta through Interference of Plane Waves":

It is immediately clear that through the superposition of a great number of plane waves with the same wave normal and closely neighboring frequencies one can produce a "signal" that is limited almost exclusively to a thin plane parallel "slice" of the total space. On the other hand, one is perhaps in doubt for a moment whether and how it is possible to restrict the signal to a small region of space in all three directions. According to Debye and Laue, one achieves this by allowing not only the frequencies to vary slightly, but also the wave-normals over a small region, a small solid angle $d\omega$, and then integrates together a continuum of infinitesimal wave functions within this range of frequencies and wave-normals.

Of course one naturally cannot guarantee through classical wave laws that the "model of a light quantum" thus produced – which after all still mixes wavelengths in every direction – will also permanently stay together. On the contrary, it will be scattered into ever greater space after passing through a focus point.

This discussion foreshadows his wave-mechanical representation of a particle by a wave packet. When he did achieve his wave equation for material waves, this spreading of the wave packet would turn out to be one of the ineluctable features of wave mechanics and the source of some of his most desperate problems in its interpretation.

CHAPTER 6

Discovery of wave mechanics

When Schrödinger returned from his holidays for the winter semester of the academic year 1925 to 1926, he would have been unaware that he was on the verge of a period of creative activity that would change forever the world of physics. Hermann Weyl once said that Schrödinger "did his great work during a late erotic outburst in his life." His marriage with Anny was at a high point of tension, with constant talk of divorce. Their Zurich circle provided ample opportunities for amorous adventures, and if Erwin cared to venture outside the usual academic liaisons, both Weyl and Debye were available as guides to the night life of the city.

BEFORE THE BREAKTHROUGH

The semester began on October 15, and Schrödinger had his usual heavy teaching schedule. The joint seminar with E.T.H. met for two hours every fortnight. Debye suggested that he give a talk on the de Broglie thesis, which had just been published in *Annales de Physique*. He said in effect, "Schrödinger, I don't understand this de Broglie business. You read it. See if you could get a nice talk about it."

On November 3, Schrödinger wrote to Einstein:

A few days ago I read with the greatest interest the ingenious thesis of Louis de Broglie, which I finally got hold of . . . The de Broglie interpretation of the quantum rules seems to me to be related to my note in *Zeit. f. Physik 12, 13, 1922*, where a remarkable property of the Weyl "gauge factor" along every quasi-period is shown. As far as I can see, the mathematical content is the same, only mine is much more formal, less elegant, and not actually shown in general. Naturally de Broglie's consideration in the framework of his comprehensive theory is altogether of far greater value than my single statement, which I did not know what to make of at first.

He was also corresponding with Alfred Landé, professor of physics at Tübingen. On November 16, he wrote:

I was especially pleased with your news that your work would be a "return to wave theory." I am also strongly inclined that way. I have been intensely concerned these days with Louis de Broglie's ingenious theory. It is extraordinarily exciting, but still has some very grave difficulties. I have tried in vain to make for myself a picture of the phase wave of the electron in the Kepler orbit. Closely neighboring Kepler ellipses are considered as "rays." This, however, gives horrible "caustics" or the like for wave fronts.

A caustic is a curve that gives the boundaries of an initially plane wave after reflection or refraction. The electron waves of the de Broglie theory are traveling waves, and Schrödinger wanted to find the structure of such waves when they are refracted sufficiently to travel in one of the Bohr orbits.

His colloquium report on the de Broglie thesis was most likely given at the meeting on December 7. According to the recollections of Felix Bloch, at the end of this colloquium:

Debye casually remarked that he thought this way of talking was rather childish. As a student of Sommerfeld he had learned that, to deal properly with waves, one had to have a wave equation . . . Just a few weeks later [Schrödinger] gave another talk in the colloquium which he started by saying: "My colleague Debye suggested that one should have a wave equation; well, I have found one!"

It is most likely that this second talk was given after the Christmas holidays. It should be noted, however, that Debye himself had no recollection of ever mentioning the need for a wave equation.

Schrödinger's first derivation of a wave equation for particles was based entirely upon the relativistic theory as given in de Broglie's thesis, but he did not publish this equation. The crucial test of any theory would be its application to the problem of the hydrogen atom. At the non-relativistic level, it must at least be able to reproduce the results of the old quantum theory of Bohr for the energy levels and quantum numbers. A relativistic theory should also be capable of explaining the Sommerfeld equation for the fine structure and perhaps improving upon it.

A considerable mystery now obscures the historical record. Schrödinger did not keep systematic, dated research notebooks; he used separate notebooks for different subjects, and entries were seldom dated. The surviving records from this time (about December 1925) are a three-page set of rough notes (which we shall call N1) called "H Atom – Characteristic Vibrations [*Eigenschwingungen*]," and a seventy-two-page research notebook (N2) titled "Eigenvalue Problem of the Atom I." N1, which is believed to be the earlier, contains the first written record of a wave equation for the hydrogen atom; it is a relativistic equation, as will be discussed below.

Schrödinger must have accomplished some of the work on the relativistic equation in November. Hans Thirring recalled that "In November 1925 Schrödinger wrote me a four-page letter in which he revealed his wave mechanics. This letter, like many others, was lost when the Gestapo seized my correspondence." This letter may have contained the relativistic wave equation for the Kepler problem shown in N1 but it is not possible that it contained the complete solutions, with the eigenvalues and eigenfunctions. A few days before Christmas Schrödinger went to Arosa, where he stayed until January 9, 1926.

CHRISTMAS AT AROSA

It is believed that Erwin wrote to "an old girlfriend in Vienna" to join him in Arosa, while Anny remained in Zurich. Efforts to establish the identity of this woman have so far been unsuccessful, since Erwin's personal diary for 1925 has disappeared.

Whenever Erwin and Anny visited Arosa they stayed at the Annex of the Villa Dr. Herwig, in a room next to the living quarters of old Dr. Otto Herwig. This is where Erwin spent part of his *Liegekur* from July 1 to October 29, 1922. The more severely ill patients lived in the main building, and Dr. Herwig was always careful to separate those less ill as much as possible from them. Erwin and Anny also stayed in the same room during the Christmas holidays in 1923 and 1924, as is attested by the hotel registry. There is, however, no record of his stay there in 1925, although he gave that address to his correspondents, so that either he stayed there without registering, or he stayed elsewhere and wished to conceal his location.

Like the dark lady who inspired Shakespeare's Sonnets, the lady of Arosa may remain forever mysterious. We know that she was not Lotte or Irene. In all likelihood she was not Felicie; her husband had lost his fortune in the post-war inflation and had gone to Brazil, leaving her with an infant daughter. Whoever may have been his companion, the increase in Erwin's powers was dramatic, and he began a twelve-month period of sustained creative activity that is without a parallel in the history of science. When he was enthralled by an important problem, he was able to achieve intense and absolute concentration, bringing to bear all his great powers as a theorist.

On December 27, he wrote from Arosa to Willy Wien in Munich.

At the moment I am struggling with a new atomic theory. If only I only knew more mathematics! I am very optimistic about this thing and expect that if I can only . . . solve it, it will be _very_ beautiful. I think I can specify a vibrating system that has as eigenfrequencies the hydrogen _term_ frequencies – and in a relatively natural way, not through _ad hoc_ assumptions . . . I hope I can report soon in a little more detailed and understandable way about the matter. At present I must learn a little more mathematics in order to survey completely the vibration problem – a linear differential equation similar to Bessel's, but less well known.

This letter shows that Schrödinger now had clearly in hand the way in which quantization conditions with whole numbers arise as eigenvalues of a wave equation. (If a wave is held fixed within definite boundaries, only certain wave patterns are allowed. These are called the *eigenfunctions* of the wave equation. They are specified by parameters such as their energies or angular momenta, which are called the *eigenvalues*.) It is highly probable that by now he had separated the three-dimensional wave equation and found how the azimuthal and magnetic quantum numbers arise from the angular parts of the equation. He has intuitively realized that the principal quantum number will arise in a similar way from the radial wave equation, but he has not yet been able to solve the latter.

The relativistic and non-relativistic radial equations are so similar that the solution of one could be followed almost immediately by the solution of the other. Since in his first paper on wave mechanics (Q1) he thanks Hermann Weyl for indicating the way to solve the equation, he would not have found the method till shortly after his return to Zurich on January 9. The actual solution would then take no more than a day or so, so that by January 11 he should have had the solution to the relativistic wave equation (now called the Klein–Gordon equation). Then followed the decision to present only the non-relativistic theory. The time schedule was rather tight but quite possible. Unfortunately, we do not have a manuscript for the earlier (relativistic) paper. The details of its content are in Notebook N2. In 1956, he wrote to Wolfgang Yourgrau:

Admittedly the Schrödinger theory, relativistically framed (without spin) gives a formal expression of the fine-structure formula of Sommerfeld, but it is incorrect owing to the appearance of half-integers instead of integers. My paper in which this is shown has . . . never been published; it was withdrawn by me and replaced by the non-relativistic treatment.

The reason why Schrödinger's relativistic equation did not agree with experiment is of course that it did not include electron spin, which was not known at that time. It is a perfectly good equation for particles of spin zero.

THE RELATIVISTIC WAVE EQUATION

The steps leading to the relativistic wave equation are sketched briefly in the three pages of Notebook N1. Schrödinger used a method similar to that found in many elementary textbooks today: substitution of the de Broglie relation for the wavelength or frequency of a particle into the ordinary steady-state wave equation of mathematical physics. The procedure is so simple that perhaps he believed it would not be convincing. When he published his equation, he gave two different and more difficult "derivations."

Of course, one cannot actually *derive* the Schrödinger wave equation from classical physics. The so-called "derivations" are therefore justifications rather than derivations. It would be reasonable simply to write down the final equation as a *postulate* of wave mechanics, and then to show that it gives correct results when applied to various calculations. But before you can write it down, you have to have it, and it did not spring fully fledged from Erwin's mind like Venus from the scallop shell. The paths leading to the equation are thus of interest, since they show the process by which the discovery was made. Schrödinger saved correspondence and notes about many uninteresting things, wrote several short autobiographical accounts, and talked on all sorts of subjects to friends of all persuasions, yet he never gave a chronological account of his pathway to the discovery that made him one of the immortals of science.

Schrödinger's wave equation was first written down in the early Notebook N1, and he worked intensively on it during his stay at Arosa. When he returned to Zurich, Dean Schlaginhausen asked him if he had enjoyed skiing during his vacation. Erwin replied that he had been distracted by a few calculations. This may have been the only time in his life when he allowed anything to distract him from a holiday.

It is not surprising that Schrödinger had difficulty in solving the radial equation. It presented an unusual type of eigenvalue problem that had not yet been discussed in the mathematical literature. Physically considered, the electron in the hydrogen atom can either be bound to the nucleus or dissociated from it. In the former case, its energy eigenvalues are discrete, but in the latter case, they have a continuous range. The discrete solutions become more and more closely packed as they converge to the dissociation point and merge into the continuum. With the help of Weyl, Schrödinger was able to solve this problem and proceed to a complete solution for the hydrogen atom.

In November 1926, he wrote a foreword to a collected edition of his first six papers on wave mechanics (which we shall denote as Q1 to Q6).

With reference to the six papers, whose present republication was caused solely by the strong demand for reprints, a young friend of mine recently said to the author: "Hey, you never even thought when you began that so much sensible stuff would come out of it." This expression, with which I fully agree after suitable discounting of the complimentary adjective, may recall the fact that the works united here in one volume emerged one after the other. The knowledge of later sections was often still unknown to the writer of the earlier ones.

The young friend was fourteen-year-old Itha Junger, whose acquaintance will be made subsequently. She remembered that when Erwin wished to work without any distraction, he used to place a pearl in each ear to shut out noise, to which he was very sensitive. During the early months of 1926, as his creativity reached a peak and the papers poured from his pen, he must have found frequent need for this stratagem.

THE FIRST PAPER

The first communication on "Quantization as an Eigenvalue Problem" (Q1) was received by the *Annalen* on Tuesday, January 27, 1926. It was less than three weeks since Erwin had returned from Arosa. In that time, he had consulted Weyl more than once, found the solution to the relativistic equation, written a paper about it, then found the non-relativistic equation, written another paper, and withdrawn the first effort because of its disagreement with the hydrogen-atom data. There is no evidence that the first paper was actually sent to the journal and no manuscript has been found.

In Q1 he gave a "derivation" of the wave equation which seems almost deliberately cryptic. He knew the equation before he devised the "derivation," and his main purpose was to show that his equation gives the correct quantization of energy levels for the hydrogen atom.

In this communication I wish first to show in the simplest case of the hydrogen atom (nonrelativistic and undistorted) that the usual rules for quantization can be replaced by another requirement, in which mention of "whole numbers" no longer occurs. Instead the integers arise in the same natural way as the integers specifying the number of nodes in a vibrating string. The new conception can be generalized, and I believe it touches the deepest meaning of the quantum rules.

The derivation used a standard procedure of mathematical physics by applying the calculus of variations to the Hamilton–Jacobi equation of classical mechanics. The final form of his wave equation is:

$$\Delta\psi + (8\pi^2 m/h^2)(E - V)\psi = 0.$$

Here ψ denotes the amplitude of the wave function of the electron, and Δ is the Laplacian operator. This paper (Q1) has been universally recognized as one of the greatest achievements of twentieth-century physics. As Dirac was to remark later, it contains much of physics and, in principle, all of chemistry. By 1960 more than 100,000 papers had appeared based on application of the Schrödinger equation. From the beginning, it was accepted as a mathematical tool of unprecedented power in dealing with problems of the structure of matter. Almost from the beginning, scientists began to ponder and worry and argue about what it might be telling them about the nature of the physical world. As J. Robert Oppenheimer wrote:

Here is this quite beautiful theory, perhaps one of the most perfect, most accurate, and most lovely man has discovered. We have external proof, but above all internal proof, that it has only a finite range, that it does not describe everything that it pretends to describe. The range is enormous, but internally the theory is telling us, "Do not take me absolutely or seriously, I have some relation to a world that you are not talking about when you are talking about me."

Schrödinger devoted several pages at the end of his paper to an interpretation of his new theory.

Naturally it is very obvious to relate the function ψ to a vibration process in the atom, to which we can ascribe more than today's doubtful reality of the electronic orbits. Originally I had the intention of establishing the new foundation of the quantum rules in this more pictorial [*anschaulich*] way, but then gave preference to the above more mathematical formulation, because it allows the important points to emerge more clearly. What seems to me to be important is that the mysterious "whole number requirement" no longer appears, but is, so to speak, traced back to an earlier stage: it has its basis in the requirement that a certain spatial function be finite and single-valued.

I don't want to go into more detail now about possible interpretations of the vibration process until it is clear that the new method can give new results. We don't know for sure that its results may not be merely a rehash of the conventional quantum theory.

He now mentions that his work has been inspired by the ingenious *Thèses* of Louis de Broglie. The difference is that the latter based his theory on traveling waves, whereas the present theory is based on standing waves. "I have recently shown that the Einstein gas theory can be derived from such standing eigenvibrations, to which the de Broglie dispersion law is applied. The above considerations for the atom can be considered as a generalization of that application to the gas model." This statement is direct evidence that the work on statistical gas theory was indeed the intellectual precursor of the discovery of wave mechanics.

The final two pages of the paper were an attempt to give a picture of the vibration processes responsible for the Bohr frequency conditions for emission and absorption of radiation. He imagines the ψ-function of the atom to be vibrating with a "pot-pourri" of very high frequencies. The frequencies given by the Bohr condition

$$hv = E_2 - E_1$$

are then visualized as beats between these vibration frequencies. He seemed pleased with this picture and remarked:

It is scarcely necessary to emphasize how much more appealing than the conception of jumping electrons would be the conception that in quantum transitions the energy passes from one vibration pattern to another. The change in the vibration pattern can take place continuously in space and time, and can readily persist as long as the emission process does.

Thus in the same paper in which he created a revolution in quantum theory and opened the way to a new era in microphysics, Schrödinger tried to use his discovery as a pathway back to a classical physics based on a continuum undisturbed by quantum jumps.

THE SECOND PAPER

Just four weeks after the first paper (Q1) the *Annalen* received on February 23 the second paper (Q2) in the series "Quantization as an Eigenvalue Problem." It consists of a detailed exploration of the Hamiltonian analogy between mechanics and optics, leading to a new "derivation" of the wave equation, an analysis of the relations between geometrical and undulatory mechanics, and applications of the wave equation to the harmonic oscillator and the diatomic molecule. The paper is a masterpiece of scientific exposition, and it must have given its author a justified feeling of satisfaction.

Schrödinger remarked that:

we know today in fact that our classical mechanics fails for very small dimensions of the path and for very great curvatures. This failure is so similar to that of geometrical optics that it becomes a question of searching for an undulatory mechanics by further working-out of the Hamiltonian analogy.

The particle of the mechanical system must thus be represented by a wave group with small dimensions in every direction (now called a *wave packet*). In such a packet one must be able to neglect any spreading of the waves in comparison with the dimensions of the path in the system. The particle corresponds to that point where a certain continuum of wave forms coalesces with the same phase. This image point (or mechanical "particle") moves with the group velocity of the wave packet. The true mechanical process is represented by the wave pattern, not by the image point. Thus:

No special meaning is to be attached to the electron path itself . . . and still less to the position of an electron in its path . . . The wave group not only fills the whole path domain all at once, but also extends far beyond in all directions . . . This contradiction is so strongly felt that it has even been doubted whether what goes on in an atom can be described within the scheme of space and time. From a philosophical standpoint, I should consider a conclusive decision in this sense as equivalent to a complete surrender. For we cannot really alter our thinking in terms of space and time, and what we cannot comprehend within it, we cannot comprehend at all. There <u>are</u> such things but I do not believe that atomic structure is one of them.

The most important advance in Q2 is the demonstration that ψ is a function in configuration space, for example, for two particles, $\psi(x_1, x_2, y_1, y_2, z_1, z_2)$. This fact establishes the relationships between the states of different particles in a system, and is the basis of all atomic and molecular structure.

REACTIONS TO Q1 AND Q2

Schrödinger sent Max Planck a reprint of Q1 and on April 2, 1926, Planck replied to thank him for the paper, which he had read "like an eager child hearing the solution to a riddle that had plagued him for a long time." Schrödinger at once replied to the "highly revered Herr Geheimrat" that the card had made him very happy, and he included a brief account of his work on the Stark effect (Q3), in which "the fundamental assumption is that the space density of electricity is given by the square of the wave function."

On May 24, Planck wrote about Q2: "You can imagine with what interest and enthusiasm I plunged into the study of this epoch-making work, although I now progress very slowly in penetrating thought processes of this kind." He was pleased to learn that the German Physical Society had invited Schrödinger to give a lecture so that the Berlin physicists would soon have an opportunity to listen to a first-hand account of this work. He then added a rather curious remark: "I do not know whether you already know Berlin, but I hope that you will find that in certain respects one lives here more freely and independently than in a smaller city, where each one checks up on the other, and one does not have the possibility of sometimes going into seclusion without anyone noticing it." Planck had already conceived the plan of bringing Schrödinger to Berlin as his successor after his retirement in 1927.

On April 16, Einstein wrote to say that Planck had shown him the wave mechanics papers and he thought that "the idea of your work springs from true genius!" No praise could have pleased Schrödinger more than these words from the scientist he admired above all others. The schoolboy who was always first in class had at last taken his rightful place at the forefront of physics. Ten days later, Einstein wrote again: "I am convinced that you have made a decisive advance with your formulation of the quantum condition, just as I am convinced that the Heisenberg–Born method is misleading."

Paul Ehrenfest wrote from Leiden on May 19: "I am simply fascinated by the . . . theory and the wonderful new viewpoints that it brings. Every day for the past two weeks our little group has been standing for hours at a time in front of the blackboard in order to train itself in all the splendid ramifications."

A more cautious note had come from Willy Wien: "How would you derive black-body radiation in your theory? . . . The physical significance of the constant h has not become clear to me . . . In any case I congratulate you on your achievement and wish you good luck with all my heart."

MATRIX AND WAVE MECHANICS

After finishing Q1 and Q2, Schrödinger had time to consider the relation of matrix mechanics to his wave mechanics. In a paper received by *Zeitschrift für Physik* in July 1925, Werner Heisenberg, then twenty-four years old, had given a preliminary account of a new and highly original approach to the mechanics of the atom. Having finished his Ph.D. with Sommerfeld in Munich, he was working with Max Born in Göttingen. He had been influenced considerably by discussions with Niels Bohr. Since the Bohr, Slater, Kramers paper, the Copenhagen physicists had been fascinated by the concept that an atom can, in a sense, be equated with a set of oscillators having frequencies equal to its absorption frequencies. Heisenberg's aim was to develop a mathematical theory based entirely on experimentally observable quantities such as the frequencies, amplitudes and polarizations of spectral lines, without any reliance upon pictorial models of atomic structure based on concepts such as "the position of an electron within the atom," which he believed to be in principle unobservable. This was a thoroughly Machian (or positivist) program.

Heisenberg proposed to substitute for the position coordinate of an electron a two-dimensional array of terms based on transitions between energy levels. As is often the case with the first expression of a new idea, his paper is somewhat obscure, but while he was away from Göttingen on a visit to Cambridge, the concepts in his paper were extended and clarified by Born and Pascual Jordan (September, 1925), and when he returned he joined these two colleagues in a definitive paper on the new quantum mechanics (November, 1925).

Born had recognized immediately that Heisenberg's arrays are infinite matrices, the algebra of which was well known. Multiplication of matrices does not obey the usual commutation law, i.e. $AB \neq BA$. The classical momentum p was also replaced by a matrix. The q and p matrices follow a commutation law,

$$\mathbf{qp} - \mathbf{pq} = (ih/2\pi)\,|\,1\,|$$

where 1 is the unit matrix. The classical Hamiltonian equations of motion were now applied to the matrix quantities, $\mathbf{q}$ and $\mathbf{p}$.

The conservation of energy then requires that H (the sum of the kinetic and potential energies) be represented by a diagonal matrix in which the diagonal elements H_{nn} give the energies of the various states of the system so that the mathematical technique of matrix mechanics is the diagonalization of a matrix.

Schrödinger was aware of the first two papers on matrix mechanics when he devised his wave mechanics, but he did not have the definitive third paper. They had no influence on his work since their approach was so different from his. He was surprised to find, however, that when the same problem was treated by both methods, the harmonic oscillator for example, they gave concordant results. Heisenberg described his work as "the true theory of a discontinuum," whereas Schrödinger's concept of a wave function filling all space evoked a continuum even more pervasive than that of classical mechanics. At first he could see no relationship between the two theories, but in the last week of February and the first two weeks of March, he was able to establish the relationship in one of his most brilliant papers. (Carl Eckart at Caltech also did this independently.)

The remarkable analysis in this paper showed that the two theories appear to be equivalent from a mathematical stand-point. Here then are two theories, one based on a clear conceptual wave model of atomic structure and the other based on a radical statement that any such model is meaningless, yet both lead to the same final results. Schrödinger was taken aback by what he had discovered:

There are today not a few physicists who, exactly in the sense of Mach and Kirchhoff, see the task of physical theory to be merely the most economical description of empirical connections between observable quantities . . . In this view, mathematical equivalence means almost the same as physical equivalence.

The spirit of Mach might consider matrix mechanics, with its complete lack of any model (*Unanschaulichkeit*) to be superior to wave mechanics. Schrödinger, however, felt "repelled" by the Heisenberg theory. He believed that to deprive a physicist of the possibility of making space-time models of subatomic phenomena would inhibit further progress.

THE THIRD PAPER

The third paper in the series, Q3, was received by *Annalen* on May 10. It is an extensive account of perturbation theory and its application to the effect of an electric field on the hydrogen spectra (Stark effect). In many instances an exact (analytic) solution of a differential equation cannot be obtained, but it is possible to obtain an approximate solution by applying a small disturbance (or perturbation) to a system for which an exact solution is available. Such methods were often used in classical theories, for instance, the treatment by Rayleigh of vibrations in a string with small inhomogeneities. In atomic theory there are many interesting cases in which more than one eigenfunction for the unperturbed system have the same eigenvalue (so-called *degenerate* eigenvalues) and a perturbation causes a splitting of these eigenvalues. In the Stark effect, the lines in an atomic spectrum are split by an applied electric field. Schrödinger calculated this effect for the hydrogen spectrum and obtained the famous Epstein formulas for the Stark displacements; the calculated intensities and polarizations were in agreement with Stark's experimental values. He considered this to be the first major application of the new theory.

CORRESPONDENCE WITH LORENTZ

Schrödinger sent reprints of Q1, 2, 3, and the paper showing the relation between wave and matrix mechanics to Hendrik Lorentz in Leiden, then seventy-three years old, the grey eminence of theoretical physics. On May 27 Lorentz wrote a long letter with some cogent criticisms. In his opinion there were many difficulties yet to be overcome in the theory. He objected particularly to the representation of a particle as a wave packet, since such a packet does not remain compact with time but gradually spreads out. In the field of the hydrogen atom, this expansion would occur rapidly, since a wave packet can persist for an appreciable time only if its dimensions are large compared to the wavelength, a condition not satisfied in the Bohr orbits of the atom. Next, he raised the question of why the field of the electron itself is omitted from the wave equation. He admitted that to include it would cause untold difficulties and spoil the calculation of the energy eigenvalues. Schrödinger's view, however, might solve this difficulty, for "if the electron as such is no longer there, then one can be more readily satisfied that only the term for the nuclear charge appears."

Schrödinger replied to Lorentz: "You have done me the extraordinary honor, in eleven closely written pages, to submit the ideas of my last works to a deeply searching analysis and critique. I can find no words to thank you sufficiently for the valuable gift that you have thereby presented to me." He then responded to some of the points raised.

He admitted a difficulty in the interpretation of the absolute value of ψ^2 as a density of charge. "What is unpleasant here," he said, "and indeed directly to be objected to, is the use of complex numbers. ψ is surely fundamentally a real function." At this time he had not realized the fundamental, and indeed epoch-making, significance of the introduction of complex numbers into the theory, as carriers of the unobservable phase information of the ψ waves.

Concerning the spreading of the wave packet, he sent a copy of his paper on the harmonic oscillator (Q6) in which a wave packet can be constructed that does not expand with time. He still thought the same sort of thing can eventually be constructed for other cases, although the problem then appeared "hopeless." In the event, this optimism was not justified, and the harmonic oscillator is a special case arising from its equidistant energy levels.

Lorentz replied on June 19. He supposed that some of his own difficulties were due to the fact that he was still involved in the thought patterns of the old quantum theory. He was pleased to get the harmonic-oscillator paper and at first he thought that the wave-packet idea might be satisfactory after all, but his joy was soon extinguished, for in the case of the hydrogen atom, one does not have available the short wavelength vibrations that would permit a wave packet to be constructed. He sent a detailed twelve-page calculation to demonstrate this fact. Schrödinger soon de-emphasized the wave-packet picture, which was not an integral part of his theory; although it helped to support his idea that waves constitute the basic reality of the subatomic world.

This exchange of letters with Lorentz was very helpful to Schrödinger, since it provided a profound yet sympathetic criticism of some of his physical ideas. From June 1926, his conviction of the primacy of wave motion as the source of physical reality began to waver. Part of his difficulty was the conflict between his philosophical beliefs and his scientific methodology. A persuasive case can be made that at this time he was a disciple of Mach in his epistemology, of Vedanta in his ontology, but a follower of Boltzmann in his scientific methodology. Thus he demanded that a scientific theory provide pictures of reality in space and time, but he did not believe that the pictures portrayed a real world. He had a much more complex and subtle view of the world than the simplistic positivism of Göttingen and Copenhagen, but in 1925–26 he was creating new science at such an unbelievable pace that he had no time to consider the philosophic implications of his own creations.

THE FOURTH PAPER

The fourth and final paper (Q_4) of this marvelous series was received by *Annalen* on June 23. So far Schrödinger had treated the wave mechanics of static systems only. Now he turned to problems in which the system is changing with time. These include scattering problems, in particular the scattering of radiation by atoms and molecules, and also the absorption and emission of radiation. In principle also, the general theory comprises chemical kinetics, although in practice the calculations have so far been feasible only for simple gas reactions.

Q_4 was the culmination of six months of research activity that is without an equal in the history of science, both in the intensity of its creativity and in the importance of its results for subsequent progress in physics and chemistry. The results were achieved by one man, working alone, except for occasional discussions with Hermann Weyl about mathematical questions, and with Erwin Fues as a young audience for new ideas and occasional helper with numerical calculations.

Up till now Schrödinger had been convinced that Ψ must be a real function. He was troubled by the $\sqrt{-1}$ factors that occurred in the theory, but he thought that they could be avoided by simply taking the real part of complex terms, so that the use of $i = \sqrt{-1}$ would be merely a convenient device for calculations such as that used in electrical-circuit analysis. Sometime between June 11 and June 21 Schrödinger arrived

at the conviction that Ψ is complex. Heisenberg had introduced $\sqrt{-1}$ in his commutation relation, $pq - qp = -ih/2\pi$,
but without any interpretation. From Schrödinger's Q4, it can
be seen that the wave function is a product of an amplitude
factor and an imaginary phase factor. As Dirac said later, the
phase quantity was very well hidden in nature and it is because
it was so well hidden that people had not thought of quantum
mechanics much earlier. The real genius of Heisenberg and
Schrödinger was to discover it. It is the source of all interference phenomena.

The Schrödinger wave equation including the time is even
more important than his equation for stationary states. It can
be written as:

$$i(h/2\pi)\,\dot{\Psi} = H\Psi$$

where H denotes the Hamiltonian operator, and $\dot{\Psi} = \partial\Psi/\partial t$.
This is the equation that was to appear on the first-day
postmark of the Austrian stamp commemorating Schrödinger's
one-hundredth anniversary.

Schrödinger applied his equation to the important problem
of the interaction of atoms with radiation. He used perturbation theory with a small time-dependent perturbation. His
theory can be called "semi-classical," since the atom is treated
as a wave mechanical system whereas the radiation field is
treated classically. (In 1927 Dirac introduced the method of
"second quantization" in which the radiation field is also
treated by quantum-mechanical theory.)

In a final section of the paper, Schrödinger returned to a
discussion of the physical significance of the wave function. His
discussion is modest and tentative; although he was later to be
accused of advocating an untenable interpretation, such an
accusation is not borne out by his actual words:

$\psi\psi^*$ is a sort of *weight function* in the configuration space of the system.
ψ^* is the complex conjugate of ψ, in which i becomes $-i$. The *wave
mechanical* configuration of the system is a *superposition* of many, strictly
speaking *all*, the kinematically possible point-mechanical configurations. Thereby every point-mechanical configuration contributes to
the true wave mechanical configuration with a certain weight, which
is given precisely by $\psi\psi^*$.

If one likes paradoxes, one can say that the system is found simultaneously in all conceivable kinematic locations but not in all of them "in equal strength" . . . That the ψ itself in general cannot be given a direct interpretation in three-dimensional space, as in the one-electron problem, because it is a function just in configuration space, not in real space, has been stated repeatedly.

This statement is remarkably similar to the statistical interpretation of the wave function as given by Max Born, who submitted his first paper on this subject just a few days after Schrödinger submitted Q4. This similarity makes all the more surprising Schrödinger's rejection of the statistical interpretation which Born gave. According to this, $\psi\psi^*$ gives the *probability* that the particle be found in a specified region of space, while the ψ waves do not represent any physical entity traveling through space.

In the last paragraph of Q4 Schrödinger remarked "that a certain difficulty no doubt still lies in the use of a complex wave function. If it was fundamentally unavoidable, and not a mere convenience for calculation, it would mean that actually two wave functions exist, which only together give conclusions about the state of the system." Apparently he did not yet fully appreciate that the complex function contains the hidden phase information essential for a complete specification of the ψ waves, but the light of this truly remarkable fact would soon dawn upon him.

THE PHYSICISTS

The last of the six major papers was sent to the journal on June 21, and with it Schrödinger's greatest scientific achievements were concluded. Such is the natural history of theoretical physicists. Schrödinger was unusual only in the relatively advanced age at which his great work was achieved. Heisenberg reached his peak at twenty-seven, Dirac at twenty-eight, Bohr at twenty-nine, Pauli at thirty-one, and Einstein at thirty-six. It should not be thought, however, that these scientists no longer made important contributions to their subject. After the research breakthroughs came the time for teaching, writing, philosophical discussions, and the inspiration of the next generation.

No sooner had Schrödinger laid down his pen at the completion of Q4 than he had an opportunity to present his ideas to an international audience. Debye and Scherrer had arranged a "Magnetic Week" in Zurich from June 21 to 26, with physicists coming from all over the world, notably Sommerfeld, Pauli, Langevin, Pierre Weiss, and Otto Stern. Besides the scientific meetings, there was an excursion on the lake with opportunities for shipboard confidences enlivened by cold white wine. Charles Mendenhall from Wisconsin was there and he invited Schrödinger to visit Madison early in 1927 for a series of lectures.

The general reaction to wave mechanics was enthusiastic, but the younger physicists especially were skeptical about any attempts to restore classical concepts of the continuum on the basis of the new theory. There were no personal hard feelings between Schrödinger and Heisenberg, but they were unsparing in their criticisms of each other's interpretation of quantum mechanics. Schrödinger felt that the lack of any pictorial model in matrix mechanics made its application to new problems practically impossible. He wrote to Lorentz that "the frequency discrepancy in the Bohr model seems to be . . . something monstrous." Heisenberg wrote to Pauli: "The more I think of the physical part of the Schrödinger theory, the more abominable I find it. What Schrödinger writes about *Anschaulichkeit* makes scarcely any sense, in other words I think it is bullshit [*Mist*]."

Pauli tended to side with Heisenberg and he once referred to the *Züricher Lokalaberglauben* (Zurich local superstitions). Erwin was upset when he heard this phrase, but Pauli wrote a soothing explanation: "Don't take it as a personal unfriendliness to you but look on the expression as my actual conviction that quantum phenomena naturally display aspects that cannot be expressed by the concepts of continuum physics (field physics). But don't think that this conviction makes life easy for me. I have already tormented myself because of it and will have to do so even more."

Erwin was appeased and wrote: "we are all nice people, and are interested only in the facts and not in whether it finally comes out the way one's self or the other fellow supposed. If outsiders, all the same, find us capricious, we know that such capriciousness serves science better than uniformity."

Schrödinger had received several invitations to lecture in Germany, and on July 11 he traveled to Stuttgart, where he stayed with the Regeners for several days, before proceeding to Berlin, where he stayed with the Plancks. His lecture before the German Physical Society on July 16 was entitled "Foundations of Atomism Based on the Theory of Waves," and on July 17 he gave a more specialized talk to the university physics colloquium, which was followed by a party that evening at the Planck house. The older generation of Berlin physicists, Einstein, Planck, Laue, and Nernst, were enthusiastic about both his mathematics and his semi-classical interpretations. Planck began to consider seriously his plans to bring Schrödinger to Berlin as his successor when he retired in the following year.

Erwin's next stop was Jena, the university where he had been a lowly assistant just five years previously. His old friends there gave him an enthusiastic welcome. Nothing makes a scientist happier than to lecture as a great man at the scene of his apprenticeship.

On July 21, in Munich, he talked at Sommerfeld's "Wednesday Colloquium" and two days later repeated his Berlin lecture for the Bavarian branch of the Physical Society. Heisenberg happened to be visiting Munich and attended the lecture. In the question time, he asked how Schrödinger ever hoped to explain quantized processes such as the photoelectric effect and black-body radiation on the basis of his continuum model. Before the speaker had a chance to reply, Willy Wien angrily broke in and, as Heisenberg reported to Pauli, "almost threw me out of the room," saying "Young man, Professor Schrödinger will certainly take care of all these questions in due time. You must understand that we are now finished with all that nonsense about quantum jumps."

Heisenberg was quite upset by the Munich seminar and the fact that he could make no impression with his arguments. He wrote immediately to Bohr to report on the situation, and Bohr wrote to Schrödinger to invite him to Copenhagen for some serious discussions.

When Erwin returned to Zurich for a belated summer vacation, he had earned the professional and personal esteem of the mandarins of German physics. He found that Anny had arranged for him an unexpected but tempting diversion, twin nymphets awaiting his instruction in high-school mathematics.

THE SALZBURG TWINS

The non-identical twin sisters, Itha and Roswitha Junger, known to their friends as Ithi and Withi, were fourteen-year-old schoolgirls from Salzburg. Their grandfather Georg Junger (1831–1908) had been a famous citizen of that city, who founded in 1858 a firm of wholesale merchants in the *Altermarkt*, which brought considerable riches to him and his two sons Hans and Carl who continued the business. Anny Bertel's mother is said to have been the illegitimate daughter of Georg Junger. The two families remained close, and Hans Junger's wife, Josefina Kohler, was Anny's godmother. The twins were born in August 1912.

During the academic year 1925–26, the twins had been in the fourth form at St. Ursula's convent school near Salzburg, and unfortunately Ithi failed her mathematics course. They were both exiled to the convent at Meinzigen, not far from Zurich, to try to make up the failed material. Here Anny met their mother, and is said to have suggested that since Erwin was competent in mathematics, he would give them some special tutoring. It was arranged that they would come once a week to the Schrödinger house for a lesson, with most of the time devoted to Ithi whose need was greater. Erwin did not know what math was prescribed for the fourth form, but he consulted Hermann Weyl, who outlined the material for the lessons. As Erwin recalled in a poem sent to Ithi in the next year,

> Auf Herrn Professor Schnitzer's Spuren
> Mit Algebra und Dreiecksfiguren

Das Ithilein zu Tode pflegt –
Das arme Kind war ganz verjägt.
Manch' anderes wäre zu berichten
Von Zürich – ich will darauf verzichtern.

On Herr Professor Schnitzer's traces / With algebra and three corn-ered races / He ran Ithi-bitti almost to death – / The poor little kid was quite out of breath. / Of Zurich there is much more could be told / But about such things I won't be so bold.

The "such things" included a fair amount of petting and cuddling, but despite these distractions, the tutorials were a success; Ithi caught up with her class and entered the fifth form at St. Ursula's. The tutor, however, had fallen in love with his pupil, and he began a patient campaign to bring about her surrender. He told her about his great discovery: "I did not write everything down at once, but kept changing here and there until finally I got the equation. When I got it, I knew I had the Nobel prize." He talked to her about religion: "I believe more in God the Father with the white beard than I believe in Nothing."

During the Christmas holidays the following year, Erwin joined the twins and their mother at Kitzbühel. Ithi was a better skier than Erwin, and the excursion was not a complete success, since he sprained his ankle. Spirits were restored by champagne for New Year's Eve.

Erwin sent Ithi a poetic birthday wish:

Als man Euch die ersten Windeln gebreitet
Hat an der Wiege die Schelle geläutet.
Hat der Narrenkönig sein Zepter geschwenkt
Und Euch Fröhlichkeit ins Leben geschenkt.

When they unfolded your diapers the very first time / The bells on your cradle did joyfully chime. / The king of the fools gave his scepter a shake / And bid you in life every happiness take.

This seems like an invitation to Freudian analysis, but as Ithi recalled, Erwin was proud of his verses, although Stefan Zweig once told him "I hope your physics is better than your poetry."

Erwin did not try too seriously to get Ithi to bed until she was sixteen. Then, one time in Salzburg, he came into her room in the middle of the night and sat on her bed and told her how much he loved and needed her. He also promised that he would take precautions to ensure that she did not become pregnant. All to no avail at that moment, but not long after her seventeenth birthday, they became lovers.

AGAINST BORN AND BOHR

After returning from his German travels, Schrödinger concentrated on the problem of how to reconcile his interpretation of continuous electron waves with phenomena such as the photoelectric effect, which seemed to demand particulate electrons and discrete light quanta. On August 25, he wrote to Willy Wien to confess his failure:

That the photoelectric effect . . . offers the greatest conceptual difficulty for the achievement of a classical theory is gradually becoming ever more evident to me. Unfortunately I can find so far . . . no solution at all to the problem, I mean I see no concrete idea or calculation that could bring one closer to understanding it. And to phantasize about it, as could perhaps be done, is in my view as easy as it is worthless . . .

But today I no longer like to assume with Born that an individual process of this kind is "absolutely random," i.e., completely undetermined. I no longer believe today that this conception (which I championed so enthusiastically four years ago) accomplishes much. From an offprint of Born's last work in the *Zeitsch. f. Phys.* I know more or less how he thinks of things: the *waves* must be strictly causally determined through field laws, the *wavefunctions* on the other hand have only the meaning of probabilities for the *actual* motions of light- or material-particles. I believe that Born thereby overlooks that . . . it would depend on the taste of the observer which he now wishes to regard as real, the particle or the guiding field. There is certainly no criterion for reality if one does not want to say: the *real* is only the complex of sense impressions, all the rest are only pictures.

Bohr's standpoint, that a space-time description is impossible, I reject *a limine*. Physics does not consist only of atomic research, science does not consist only of physics, and life does not consist only of science. The aim of atomic research is to fit our empirical knowledge concerning it into our other thinking. All of this other thinking, so far as it concerns the outer world, is active in space and time.

If it cannot be fitted into space and time, then it fails in its whole aim and one does not know what purpose it really serves.

Reading the first of these last two remarkable paragraphs, one must ask why Schrödinger has abandoned the idea that the underlying laws of physics are statistical, with pure chance and chaos at the heart of nature. Of course, Born's concept of "probability" is not the same as that used by Maxwell and Boltzmann in the kinetic theory of gases, and it would soon be shown to have some strange properties indeed. Nevertheless, one might have expected Schrödinger to welcome the Born interpretation as a victory for the ideas of Franz Exner, especially since he had himself recently suggested that $\psi\psi^*$ is a sort of "weighting function." The reason for his apostasy from the doctrine of pure chance may have been his own discovery of wave mechanics, which appeared to offer a renewal of confidence in the kind of causal, deterministic laws found in Maxwellian electrodynamics and Einsteinian gravitation. It is ironic that by choosing to be a conservative in physics after the revolution of 1926, Schrödinger joined the minority who dared to dissent from an orthodoxy based on the Born probability concept, which became known as the Copenhagen Interpretation.

COPENHAGEN

In response to a cordial invitation from Bohr, Schrödinger traveled to Copenhagen at the end of September. The most detailed account of his visit is given by Heisenberg in his remarkable book *Der Teil und das Ganze* (The Part and the Whole):

The discussion between Bohr and Schrödinger began at the railway station in Copenhagen and was carried on every day from early morning till late at night. Schrödinger lived at Bohr's house so that even external circumstances allowed scarcely any interruptions of the talks. And although Bohr as a rule was especially kind and considerate in relations with people, he appeared to me now like a relentless fanatic, who was not prepared to concede a single point to his interlocutor or to allow him the slightest lack of precision. It will scarcely be possible to reproduce how passionately the discussion was carried on from both sides.

Schrödinger contended that the idea of quantum jumps was nonsense. Transitions must occur smoothly and continuously in accord with the laws of electrodynamics. Bohr replied that the jumps did occur but we cannot visualize them according to any model based on the old physics that we use to describe the events of everyday life; the processes with which we are concerned here cannot be the subject of direct experience and our concepts do not apply to them.

Schrödinger replied that all we need do is to change the picture and say that there are no electrons as particles but rather electron waves or matter waves. The radiation of light by atoms then becomes as easy to understand as the emission of radio waves by an antenna, and the former unsolvable contradictions disappear. No, said Bohr, they do not disappear, they are simply shifted to another place. For example, to derive the Planck radiation law, it is essential that the energy of the atom have discrete values and change discontinuously.

Finally Schrödinger threw up his hands and said, "If we are going to have to put up with these damn quantum jumps, I am sorry that I ever had anything to do with quantum theory." Bohr tried to soothe him: "But the rest of us are very thankful for it – that you have – and your wave mechanics in its mathematical clarity and simplicity is a gigantic progress over the previous form of quantum mechanics."

The discussion went on in this way day and night, without reaching any agreement. After a few days, Erwin became ill with a feverish cold. Mrs. Bohr took care of him and brought tea and cake to his bedside. But Niels sat on the edge of the bed and continued the argument, "But surely Schrödinger, you must see." But Erwin did not see, and indeed never did see, why it was necessary or how it was possible to destroy the space–time description of atomic processes.

The conversations, however, deeply affected both men. Schrödinger recognized the necessity of admitting both waves and particles, but he never devised a comprehensive interpretation of quantum phenomena to rival the Copenhagen orthodoxy. He was content to remain a critical unbeliever. Heisenberg began the analysis that soon led to his principle of indeterminacy, and with this as a foundation, Bohr ventured more deeply into philosophical waters and emerged with his concept of complementarity.

After his return to Zurich, Schrödinger reported in a letter to Wien: "Bohr's . . . approach to atomic problems . . . is really remarkable. He is completely convinced that any understanding in the usual sense of the word is impossible. Therefore the conversation is almost immediately driven into philosophical questions, and soon you no longer know whether you really take the position he is attacking, or whether you really must attack the position that he is defending."

Nevertheless Erwin was deeply impressed by the spirit of Niels Bohr, and wrote to him a rather ornate letter of appreciation:

The lovely, sunny, hospitable home with its kindly people, which received a stranger like me as an old friend, surrounding me with every comfort, was an experience that the heart can never forget. But also this city, this home, this family – they are those of the great Niels Bohr; he is the one I have to thank for all this kindness through which I could speak with him for hours about those things that are so close to my heart, and hear in his own words about the positions he takes toward the many attempts to build one stage further upon the broad substantial foundation that he has given to modern physics. That was for a physicist . . . a truly unforgettable experience.

Schrödinger then said that despite the firmness with which he had advanced his own views, the psychological effect of the arguments of Bohr had been serious indeed and he was now more aware of the unresolved problems in his attempts to provide a continuous picture of the interaction of radiation with atoms, but he still believed this process could be *visualized* in space and time. Bohr did not choose to enter a detailed debate by correspondence, but he called Schrödinger's attention to recent work in Copenhagen by Oskar Klein: "If you are not able to kill off the ghosts completely in ordinary space and time, perhaps you can achieve a settlement in five dimensions."

CHAPTER 7

Berlin

Schrödinger had accepted an invitation from the University of Wisconsin to give a course of lectures early in 1927; they would pay $2,500, which included an allowance for travel costs. Shortly before he left for the United States he heard that he was a leading candidate for the succession to Max Planck at the University of Berlin.

AMERICAN VOYAGE

On December 16, Erwin and Anny attended a Christmas and Farewell Party held in the Physics Institute of the University. Edgar Meyer recited one of the long doggerel poems for which he was famous. It included the verses:

> Schon Galileo hat es uns gezeigt
> Das jeder Körper in Ruhe bleibt
> Zwingt ihn nicht eine auss're Kraft
> Zu ändern die Bewegungseigenschaft.

> Und so auch hier; denn glaubt Ihr lieben Leute
> Wir könnten Abschied feiern heute?
> Hatt' nicht die Anny zart getrieben,
> Der Erwin war zu Haus geblieben.

Galileo showed us long ago / That every body stays just so / Unless it's pushed by some outer force / To change its predetermined course.

And so dear friends are you impressed / That we gather here for a farewell fest? / For if Anny had not so tenderly nudged / Erwin from home would never have budged.

They set out on December 18 for the new world. At Le Havre they boarded the *de Grasse* for a ten-day voyage to New York. Erwin grumbled from the beginning. He found his fellow passengers distinctly unattractive "examples of the modern 'society' that I usually manage to keep at arm's distance." At dinner he was seated between two painted and powdered ladies "beyond the canonical age" and he found the "hard, ruthless expressions" of their consorts equally repulsive. Their French manners, for him, made the bad company even worse. His disposition was not helped by the fact that he was cooped up in a small cabin with Anny, who was seasick throughout the voyage.

The sight of the Statue of Liberty did nothing to alleviate his spleen: he thought this monument to be "grotesque, between the comic and the ghastly," and needing only a giant wrist-watch on the upraised arm to complete the picture. They were met at the dock by Karl Herzfeld, who had come from Baltimore to welcome them. Erwin was so distressed by the noise and dirt of New York that he threatened to take the next ship home, but Anny persuaded him to stay. He was then so frustrated by his attempt to arrange train tickets to Madison that he became nearly hysterical until Anny took over, but the comfortable drawing room and attentions of the cheerful black porter began to restore his equilibrium. He had spent only twenty-six hours in New York, but he found the experience of the city "shattering." Chicago was even worse as there was the added fear of being shot by "bandits who spring with loaded guns from speeding autos."

Madison was a great relief. Here at last was "a real city" in the European sense. They received a warm welcome from the physicists and on Sunday night attended a reception and buffet dinner for eighty persons at the Mendenhall home. Erwin was particularly pleased to meet young John van Vleck; he found that the depth of his ideas surpassed his ability to express them, "just the opposite of the situation with most women."

When Erwin decided to buy Anny a fur coat, he was amazed by the informality of the salesman, who treated him like an old friend. Unfortunately, they did not find a fit for Anny, and the salesman apologized with "Well, you see, you are a pretty big girl."

At Madison, Schrödinger's lectures were, as usual, an intellectual banquet, and he was offered a permanent professorship. He was not at all tempted by an American position, and he declined on the basis of a possible commitment to Berlin. While in the Midwest, he lectured also at the Universities of Chicago, Iowa, and Minnesota. On February 14, the Schrödingers set out for Pasadena on the California Limited. They saw the Grand Canyon and visited some Indian reservations, where Anny was dismayed by the poverty and ill health of the Indians, while Erwin was fascinated by their ceramics and textiles.

Anny found Pasadena to be "unbelievably beautiful, like a great garden." Even Erwin felt cheerful. He wished only that California were inhabited by Italians and Spaniards, instead of Americans. On their first evening, there was a party at the Millikans, and everyone was so kind and welcoming that Erwin had to admit that Americans were considerate of others to a degree quite unknown among Germans, especially northern ones.

The grand old man of theoretical physics, Hendrik Lorentz, was also in Pasadena at this time. He lectured at 3:45 p.m., after which there was a cup of tea, and Schrödinger followed at 5:00 p.m. The two Nobel prize winners, Lorentz and Millikan, sat in the front row. The California Institute of Technology was a major center of American physics, and there was a knowledgeable audience of about sixty, including all the professors of physics and physical chemistry.

Among the professors was Paul Epstein, whom Erwin called "one of the kindest, most unassuming, most worthy companions I have ever known." Paul took them to visit the beach at Santa Monica, the movie studios in Hollywood, and later on a drive up the old Mount Wilson Road. Erwin was terrified by the precipitous roadsides, but Anny enjoyed the climb, and the famous view of the lights of Pasadena and Los Angeles from the summit.

They were both sorry to leave Pasadena, and Erwin wished

they had stayed six weeks in warm California and only two in the cold Midwest. Their return trip took them through Salt Lake City, where he listened with interest to the story of the Mormons and their devotion to polygamy. There was a stop for a lecture at Ann Arbor. They stayed at the University Club, where women were admitted only by the back door. A week in Boston included three lectures at M.I.T. and three at Harvard, and a party given by the Bridgmans at a country inn.

Thence to Baltimore where Schrödinger was happy to meet Robert Wood, the great spectroscopist. He was much impressed by Wood's researches on fluorescence, and in October, 1927, nominated him for a Nobel prize in physics (which, however, he never received). They saw Washington bedecked in cherry blossoms, and stayed one night in the villa of the millionaire Edward Loomis, a great benefactor of physics at Johns Hopkins University. Through the efforts of Herzfeld, Schrödinger was offered a permanent professorship at the excellent salary of $10,000 a year, but he did not consider it seriously as an alternative to the Berlin chair.

An important factor in Erwin's intense dislike of the American way of life was "the great experiment." An occasional glass of good beer or bottle of fine wine would have made everything more bearable. With a plate of succulent Chesapeake Bay oysters he was offered a choice between sweet ginger ale and chlorinated icewater. "To the devil with prohibition," he exclaimed.

After a final lecture at Columbia, they embarked for home on the SS *Hamburg*. It had been a strenuous journey, including over fifty lectures in three months. They arrived back in Zurich on April 10, and almost immediately were faced with a decision about a move to Berlin.

THE BERLIN CHAIR

In May 1926, the Prussian Minister for Science, Art and Education asked the Philosophical Faculty of the University of Berlin to consider the future of the chair of theoretical physics, and a committee was named for this purpose. They first considered the two distinguished theoretical physicists already holding Berlin professorships, Laue and Einstein. They could see no reason to appoint Laue, since this would simply change one vacancy for another, and Einstein did not wish to accept a position that would require lecturing. It was thus necessary to "cast a glance outside."

Their first choice would be Arnold Sommerfeld, holder of the chair at Munich. He was fifty-nine years old, probably the most proficient mathematician of all the physicists; he was not only a leader in research but also the best teacher of advanced students, graduates from his department filling many of the most important positions in the world of theoretical physics. The other leading theoreticians, listed in order of seniority, were Max Born, Pieter Debye, Erwin Schrödinger, and Werner Heisenberg. Debye was not primarily concerned with the mathematical analysis of problems at the frontiers of pure physics. Thus they eliminated his name. They recognized the genius of Heisenberg, "at present the most outstanding theoretical physicist in Germany," but believed that a call to the Berlin chair at the age of twenty-three would not be in his own best interests.

Thus Born and Schrödinger remained. They found it difficult to rank these two, but Born had directed an Experimental Institute in Göttingen, and such an experience was not entirely in his favor. Regarding Schrödinger they wrote:

For some years already he has been favorably known through his versatile, vigorously powerful, and at the same time very profound style in seeking new physical problems that interested him and illuminating them through deep and original ideas, with the entire set of techniques, which mathematical and physical methods presently provide . . . In lecturing as in discussions Schrödinger has a superb style, marked by simplicity and precision, the impressiveness of which is further emphasized by the charming temperament of a South German.

Thus the final recommendation to the ministry was: (1) Sommerfeld, (2) Schrödinger, (3) Born. When Sommerfeld declined to leave Munich, Schrödinger became the leading candidate for the Planck chair.

FROM ZURICH TO BERLIN

Soon after the beginning of the summer term in Zurich, Schrödinger received a formal invitation to accept the chair in Berlin. Despite the prestige of the position, it was not easy for him to leave Switzerland, for he had grown to love the stable and peaceful life and the nearness to the high mountains, and he was aware of the vicious forces lurking behind the democratic facade of Prussian society. As soon as he informed the Dean of the Berlin offer, the University made every effort to keep him in Zurich. The Berlin salary was much higher than the maximum in a Swiss university; as Anny put it, "We got more and more and more – it was really very satisfactory." The basic annual salary was RM 13,702 plus a cost-of-living allowance of RM 3,000 and a guarantee of at least RM 10,000 for lecturing fees, making a total of RM 26,702 ($6,676). The Swiss were able to arrange a joint professorship at the E.T.H., so that the combined salaries came close to the Berlin level, but the two professorships would have required almost twice the amount of teaching.

It was usual in Switzerland to publish the news of outside offers to professors, so that the students became aware of the possibility that Schrödinger might be leaving. They organized a torchlight parade that marched through the streets around the university to the courtyard of the Schrödinger house. Such parades were traditional student demonstrations for favorite teachers, but they were quite rare, and Erwin was deeply moved.

He finally decided to accept the Berlin offer. An important reason for his decision was the personal persuasion of Max Planck. After he moved to Berlin, he wrote a heartfelt explanation in the form of a poem in Planck's guest book:

> War's nicht um Ruhm, das will ich wohl beschwören.
> Und auch – ich sag es nur mit leiser Scham –
> Auch Gold allein nicht könnte mich betören.
> Den Ausschlag gab ein Wort – aus langen Reihen
> Von Briefen, von Gesprächen, bunt und kraus,
> Verehrungswürdige Lippen sprachen's aus,
> Nicht drängend zwar. Ganz kurz: mich tät es freuen.

'Twas not for fame, which I will quite forswear, / And also – I say it only with faint shame – / Nor gold alone could now persuade me. / The scales were tipped with just one word – / From all the messages, various and intricate, / Respected lips expressed it thus, / Not pressing. Quite brief: "I for one would be glad."

Before they moved to Berlin, Erwin remarked to Anny, "You know, I have the feeling that we shall not be staying there for so very long." This was an uncanny prediction, for the Berlin chair was the apex of European physics, but he may have been thinking merely of an ultimate return to Vienna. They rented the spacious first floor of a substantial house in Grunewald (44 Cunostrasse), a western suburb adjacent to Dahlem where the Plancks lived. It was about twenty minutes by subway from the university.

The move to Berlin was completed in time for Erwin's fortieth birthday on August 12, 1927. He wrote for himself an "Epitaph if I should die today":

> Er hatte mit seinem vierzig Jahren
> Von Leben weniger erfahren
> Als manche von den Jüngern um ihn her.
> Und dennoch wusste er erheblich mehr davon,
> Wie dieses Weltgetrieb im Innersten zusammenblieb
> Als er zu sagen sich erkühnte,
> Obwohl er nicht den Namen "Prud" verdiente.
> Sein Wissen starb zum Glück mit ihm,
> Jetzt teilt er's mit den Cherubim.
> Ob denen Neues brachte sein Bericht,
> Das wusste er zu Lebzeit selbst noch nicht.

In the forty years that his course had run / He'd experienced less than many had done / Among the apostles around him here. / And yet he knew a great deal more / Of the ways the world is held together / To which he had never dared allude / Though he hardly deserved to be called a "prude." / It's lucky this knowledge died with him, / He can share it now with the Cherubim. / But whether his news would be a surprise / While still on earth he could never surmise.

Before beginning his duties in Berlin, Schrödinger traveled to Brussels to participate in what would be the most historic of the Solvay conferences on physics.

The Fifth Solvay Conference met from October 24 to 29, 1927, under the presidency of Hendrik Lorentz. This was the last public appearance of this revered elder statesman of physics, for he died a few months later. Unlike his situation at the previous conference, Schrödinger was now recognized as one of the world leaders of theoretical physics and he had been invited to give one of the "thousand-dollar lectures." The topic for the conference was "Electrons and Photons" and it was intended to provide a searching discussion of all the implications of the revolution in physics associated with the recent great advances in quantum theory.

Schrödinger's lecture was on "Wave Mechanics." He explained the importance of his time-dependent wave equation for the understanding of spectroscopic transitions. He then emphasized that, except for the case of a single particle, the wave function Ψ is not a wave in ordinary three-dimensional space. He asked "what in reality is the aspect in a space of three dimensions of the system described by this function Ψ?" His viewpoint, "perhaps a little naive," was that

the classical system of material points does not really exist, but something exists that fills all of space continuously, from which one may obtain an instantaneous "photograph" . . . if one makes the classical system pass through all its configurations, letting the image in space sojourn in each element of volume $d\tau$ for a time proportional to the instantaneous value of . . . $\Psi\Psi^*$. In other words, the real system is a composite image of the classical system in all its possible states, obtained by using $\Psi\Psi^*$ as a weighting function.

This interpretation, which he had originally put forward in Q4, differs from Born's probability interpretation in that it pictures a continuous cloud of "something" (presumably charge and mass) whereas the latter assumes point particles (electrons) within the atom. Schrödinger recognized that the continuous charge cloud must have contributions assignable to individual electrons. He admitted that "pure field theory does not suffice . . . it must be completed by a kind of *individualization* of the electric densities coming from the various point charges of the classical theory, but . . . each individual can be distributed through all space in such a way that the individuals interpenetrate."

Schrödinger's paper aroused considerable debate, and Born and Heisenberg attacked it quite vehemently. Schrödinger said it was nonsense to talk about the trajectory of an electron inside an atom, but Born said "it is not nonsense," and there this matter rested.

Einstein took no part in this particular discussion, but he did present an extended critique of the Copenhagen–Göttingen interpretation of quantum mechanics, and his debate with Bohr continued inside and outside the conference rooms. It provided the greatest excitement of the meeting and was an historic occasion, a battle of titans over the epistemological foundations of physics and over the way in which scientists should understand their world. Einstein saw immediately that the Born–Heisenberg–Bohr interpretation might conflict with special relativity by requiring that the reduction of the wave packet (localization of the electron) at one point must instantaneously prevent its reduction at an arbitrarily distant point. When the meeting ended, however, most of the physicists departed with the belief that the positivist Copenhagen view had prevailed, a belief nourished by the anti-realist philosophical tradition of central Europe. But Einstein, de Broglie, and Schrödinger were not convinced, and they left what Einstein once called "the witches sabbath at Brussels" with a resolve to fight another day.

UNIVERSITY OF BERLIN

Schrödinger's appointment as *Ordinarius* in Physics and Director of the Institute for Theoretical Physics dated from October 1, 1927, and the lectures for the winter semester of 1927–28 began on November 1, continuing until the end of February. The summer semester began in May and continued through July. Schrödinger lectured four times a week, two-hour sessions on two afternoons. In principle, his lectures were supposed to cover systematically the whole of theoretical physics in a three-year sequence and then to begin again. In the six years that he lectured in Berlin, however, no course appears to have been repeated exactly. His lecture topics were as follows: (1) Mechanics (2) Mechanics of Deformable Bodies (3) Statistical Theory of Matter and Radiation (4) Electron Theory I & II (Electromagnetic theory) (5) General Theory and Principles of Physics (6) Introduction to Quantum Theory (7) Field Physics I & II (8) Particle Physics I & II (9) Quantum Theory.

During this time, the physics faculty of the University of Berlin comprised an extraordinary group of renowned scientists and also some of the most brilliant younger European theoreticians. Nernst lectured on experimental physics, Laue on various topics from thermodynamics to optics, Lise Meitner on nuclear physics and radioactivity, Fritz London on applications of quantum mechanics including chemical bonding, Ladenburg and Paschen on spectroscopic subjects. From the winter of 1931, *Dozent* Friedrich Möglich also gave full lecture courses in theoretical topics – the students naturally called him "Unmöglich."

Schrödinger's lectures were the best – delivered in his beautifully expressive language they have been described as "always a pleasure and a delight." Thus the students were pleased with him as a teacher. He did not, however, encourage them to work under his direction on their research problems. He once said, "I am very busy, and so many research students want to come and study with me, and they ask for advice what to do. I'll tell you what I say to these students: first year do nothing but mathematics, second year nothing but mathematics, in the third year you can come and talk with me." He was acknowledged to be "a grand mathematician."

Schrödinger introduced a style of informality into his lectures that was unlike anything ever seen before at the University of Berlin, where class lines were rigidly drawn. Professors were expected to appear in a formal suit with a high white collar and tie, but Erwin usually wore a sweater and bow-tie in winter and an open short-sleeve shirt on warm days in summer. On one occasion the guard at the university gate would not let him enter until a student came and testified that he was indeed their professor.

For Schrödinger, however, as for most of the physicists in Berlin, the greatest intellectual delight was the weekly colloquium, which met on Wednesday afternoons to discuss new discoveries and theories. Einstein took a leading role, as with careful questioning and explanations he sought to reach the heart of every problem presented. Schrödinger was a less persistent questioner, often preferring only to listen until he could make an important contribution, that is, he was less concerned that Einstein to make sure that all participants understood everything. The discussion after a paper was usually longer than the paper itself, and after an hour of this, the meeting adjourned to a local tavern where a special room was reserved. The Berlin physics colloquium was a happy recompense for the spartan facilities of the university buildings. Its only problem may have been too great a uniformity in the philosophical views represented – too many conservative thinkers and too few revolutionaries willing to shake the foundations of classical physics. This conservativism was also enshrined in the powerful Prussian Academy of Sciences.

THE PRUSSIAN ACADEMY

The Academy was founded in 1700 by Gottfried Wilhelm Leibniz. Since that time its members have included many of the most eminent scholars in Europe. The membership was divided into a Physical-Mathematical Class, which included all the sciences, and a Philosophical–Historical Class, which also included classical studies and philology.

When Schrödinger came to Berlin, as for many years previously, the membership was restricted to thirty-five ordinary members in each class; in addition there were provisions for a hundred corresponding and foreign members and a few honorary members. Membership in the Academy was a great honor, and many distinguished professors in Berlin never gained admittance. The half-dozen or so Academicians in physics exerted an inordinate influence on the development of their field in Germany, since the Academy was at the apex of a system of interlocking directorates of many scientific institutions. Thus the physicists of the Academy, Planck, Nernst,

Paschen, Warburg, and Laue, were in positions of unusual power and responsibility. Einstein was also a member, and he held a special Academy professorship, but he took little part in the discussions of science policy.

At its meeting in February 1929, the Academy elected Schrödinger to membership. His inauguration as an Academician took place at a formal ceremony on July 4, 1929. At forty-two, he became the youngest member of the august society. In his short address, he paid tribute to Boltzmann and to his teacher Hasenöhrl. Of the former he said: "His ideas played for me the role of a scientific young love, no other has ever impressed me so much, no other will ever do so again." He then showed how classical mechanics is simply a first approximation to a more fundamental theory. "One of the most burning questions . . . is whether we must give up, as well as classical mechanics, its basic concept that the individual case is uniquely determined by firm laws combined with initial conditions. It is a question of the suitability of an infallible postulate of causality." The statistical interpretation of the wave function makes the properties of individual particles a matter of pure chance. He recalled his teacher Exner, who was the first to advocate an acausal interpretation of nature. He concluded, however, that there is scarcely any way imaginable to decide experimentally whether natural laws in reality are absolutely determined or whether they are partly indeterminate.

In reply, Max Planck, the Secretary of the Academy, said that he thought that Schrödinger himself, despite the neutral stance just taken, had helped to restore a more classical foundation for physics. He would like to champion a strictly causal physics. "May you, my honored colleague, in forging ahead further along the road that you have pioneered, achieve many beautiful successes. That is the sincere and optimistic good wish with which I welcome you today in the name of the Academy."

THE GOLDEN TWENTIES

The Schrödingers moved to Berlin toward the end of the "golden twenties." The German economy, financed to a large extent by American loans, was prospering and the burdens of the vindictive Versailles Treaty were less oppressive. In 1925 Field Marshal Hindenburg had been elected president, with the approval of the exiled Kaiser and even the support of the socialist party. He was a slow-witted man devoid of political ideas and distrustful of democracy, but he personified the past glory of "an army that had never been vanquished" and provided an authoritarian anchor amid the shifting currents of unstable political coalitions. As he became more senile, he was easily swayed by unscrupulous politicians like Franz Papen, who ultimately used him to transfer power to the real dictator for whom so many Germans yearned.

Planck's remark to Schrödinger that in Berlin one could do anything without being troubled by censorious neighbors was indeed true. During the twenties it was the most licentious city in Europe. Politicians and financiers set the tone, and every night they could be observed in the cabarets along the Kurfürstendamm ogling the naked women or making drunken love to young sailors. The bitter cartoons of Georg Grosz were drawn from life.

Yet, along with the pornography, there were great artistic achievements in the theater and cinema. Max Reinhardt mounted productions in two theaters, *Das Grosses Schauspielhaus*, a vast circus arena transformed into a theater in the form of a huge cave, and the more intimate *Kammerspiele*, where Marlene Dietrich made her debut in Somerset Maugham's *The Circle*. In 1930, as Lola-Lola in *Der Blaue Engel*, Dietrich portrayed the sexual ambiguity of the Prussian spirit, with its mixture of sensuality and sadism, which were to find full expression a few years later in Hitler's private army, the storm troopers of Ernst Röhm. The leading actress in the Berlin theater was Elisabeth Bergner, one of the brilliant German Jews who inspired the artistic and musical life of the city. Viennese theater was represented by Schnitzler's *Reigen*, which enjoyed an enormous success during its extended prosecution as an affront to public decency. The "proletarian theater" also offered some famous productions, especially *Die Dreigroschen Oper* of Bertolt Brecht and Kurt Weill, with Weill's wife Lotte Lenya as Jenny Diver, which opened at the *Theater am Schiff-*

bauerdamm at the end of August 1928. Brecht supported the communist party (KPD) but he became a millionaire with this commercial success. His refrain, *Erst kommen das Fressen, dann kommt die Moral* ("First comes feeding, then comes morality,") expressed the feelings of all levels of Berlin society.

Erwin was a lover of theater in Berlin as he had been in Zurich and Vienna. He did not make friends easily, especially outside the circle of his professional colleagues. Except for his fellow student Fränzel, he never in all his life had a close personal male friend, and he never had the experience of brothers or sisters or of children whom he could call his own, nor even of students for whom he could be an intellectual father and mentor. Perhaps he loved the theater because it provided vicarious experience of a gamut of personal relationships missing from his own life. In his love affairs he would act out romantic scenarios that sometimes seemed theatrical, and which probably did not always convince even their author, although it must be admitted that his heroines were not idealized with imaginary virtues, but loved with a certain scientific objectivity. At this time, Ithi Junger was his principal romantic heroine, and he would meet her in his travels away from Berlin as she did not come to visit him there until 1932. There were other loves of whom little is known.

Late in 1929, Erwin visited Innsbruck to give a lecture and he stayed with Arthur March, who had married in July. On his return to Berlin, he told Anny how much he was impressed by the beauty and charm of Hildegunde, Arthur's young wife. Hilde was a tall, slim brunette, from an unpretentious family. She was devoted to skiing, even though she walked with a slight limp due to a bad fall on the slopes. She was enthusiastic about amateur theatricals, and her generally optimistic temperament was similar to that of Erwin. Like the majority of her sex at the time, she was unencumbered with much formal education. She was to become one of the great loves of Erwin's life, but it was not love at first sight, rather more like love held in reserve, for Hilde and Arthur were presumably still in a state of connubial bliss, and the bride would have no immediate romantic interest in the distinguished Berlin professor.

Schrödinger and Einstein became good friends in Berlin. Both of them were averse to social pretensions and they shared a dislike for the formality and stiffness of the Prussian professors. Erwin loved to visit Einstein at his summer house in Caputh on the Schwielow Lake near Potsdam, and the two friends spent many hours together walking in the woods or sailing on the reaches of the Havelgewassern. Erwin had learned to sail on Lake Zurich, and he and Anny kept a small sailboat on the Wannsee.

The social life of Anny and Erwin was paradoxical. They would have happy parties for their many friends, and became quite famous for their *Wiener Würstelabende* (Viennese sausage nights), which spanned the generations from distinguished elders like the Plancks and Laues to brilliant youngsters like Max Delbrück, Viktor Weisskopf, and Eugene Wigner. When the guests were not there, however, the house was not so cheerful, as Anny fretted about Erwin's love affairs. After a lecture one afternoon in the winter of 1930, Walter Elsasser asked Schrödinger if he could discuss some points with him. "Come along to my house," Erwin said, "We can talk better there, I don't like my office here." They traveled out to Cunostrasse. Every room in the flat was brightly lighted, but Mrs. Schrödinger was not there. She came in later and disappeared almost immediately into her room at the back. After they had talked intensively for about an hour, Erwin said "Are you getting hungry?" Then they went to the kitchen and found some bread and cheese and beer, over which they continued their discussions till quite late at night. Such visits were repeated several times in the course of the next few months, always in the same way. Elsasser concluded that the Schrödinger marriage was not a happy one.

THE REVOLUTION IN PHYSICAL CONCEPTS

In May 1930, Schrödinger traveled to Munich to visit his old friend Sommerfeld and to deliver a lecture at the *Deutsches Museum*. His subject was "The Transformation of the Physical Concept of the World." Although it was not published until

1962, it is an important statement of his ideas in these early years of quantum mechanics, displaying a view that is surprisingly favorable to the Copenhagen–Göttingen interpretations, more so than most of his subsequent statements. As he once admitted to Pauli, the strenuous effort of Bohr to convert him had almost succeeded, even though he later relapsed. Thus he said:

The different wave forms, the old long-familiar electromagnetic waves as well as the new so-called matter waves, are not to be considered as purely objective descriptions of reality . . . The wave functions do not describe Nature in itself, but the *knowledge* that we possess at any given time of the observations actually carried out. They allow us to predict the results of future observations not with certainty and precision but with just that degree of unsharpness and probability with which observations actually made on the object permit predictions about it . . .

Most of us today feel that this necessary abandonment of a purely objective description of Nature is a profound change in the physical concept of the world. We feel it as a painful limitation of our right to truth and clarity, that our symbols and formulas and the pictures connected with them do not represent an object independent of the observer but only the relation of subject to object. But is this relation not basically the one true reality that we know? Is it not sufficient that it finds a solid, clear, unequivocal expression, wherein in fact all truth exists? Why must we exclude ourselves completely? Has not God himself through the mouth of our poet let it be said that we ourselves are that which must bring order out of chaotic nature? (Goethe, *Faust*)

> Das Werdende, das ewig wirkt und lebt
> Umfasst Euch mit der Liebe holden Schranken
> Und was in schwankenden Erscheinung schwebt
> Befestiget mit dauernden Gedanken.

The Becoming that eternally acts and lives / Encloses you with gentle bonds of love / And what hovers in wavering appearance / It makes secure with thoughts that last forever.

In this summary of his philosophy of science, Schrödinger seems to have met Bohr and Heisenberg more than half way. The important difference is that he still maintains the reality of the objects of physical investigation. Thus his philosophy is close to that called representational realism. It also accepts at least tacitly the duality of minds and body, subject and object, but the ontological status of the "object" is somewhat changed in that the concept of object can no longer be employed without consideration of the subject who has knowledge of the object. Schrödinger accepted this dualistic ontology only as a sort of working method for scientific research. At the deepest level, he never departed from his belief in the intuitive knowledge incorporated in the monistic doctrines of Vedanta.

The English science writer J. W. N. Sullivan included Schrödinger in a series of "Interviews with Great Scientists," published in *The Observer* in 1931. His reply to a question about the meaning of life in the universe was:

Life here is confined to a very small space in the universe. It may also be confined to a very small time . . . I am no more frightened by time than I am by space . . . If this life is the only life, then the whole meaning of the universe, throughout its extent and throughout its history, is to be found here. Although I think that life may be the result of an accident, I do not think that of consciousness. Consciousness cannot be accounted for in physical terms. For consciousness is absolutely fundamental. It cannot be accounted for in terms of anything else.

ITHI IN BERLIN

Everyone may judge for himself or herself whether or not the ethical principles that Erwin derived from Vedanta were rigorously applied in his relations with women. During these Berlin years he would meet Ithi Junger whenever possible, in Salzburg or on skiing holidays. His wooing of Ithi led him to some spectroscopic reflections on sexual psychology:

Comparable in some way to the end of the spectrum, which in its deepest violet shows a tendency towards purple and red, it seems to be the usual thing that men of strong, genuine intellectuality are immensely attracted only by women who, forming the very beginning of the intellectual series, are as nearly connected to the preferred springs of nature as they themselves. Nothing intermediate will do, since no woman will ever approach nearer to genius by intellectual education than some unintellectual ones do by birth so to speak. It has often been said that no woman really has genius. The fact is that they all do. Genius is nothing outstanding in them, it is the rule, but it is usually too weak to withstand the contamination by culture and civilization. I am fairly convinced that the only sensible thing in this moment is to make [Ithi] give herself to me completely, and to make this not by any means, but by the means of real genuine love only. If I don't succeed I will take it to be the right thing that I don't, and I will not be blamed as a shy lover, but I am fairly sure that I will succeed.

By the middle of August, she was his mistress, and Erwin was becoming more and more convinced that Ithi possessed all the qualities of a dear mistress but none of a good wife. He thought, however, that this was quite the natural thing with a girl aged seventeen and that the qualities of a "married wife" develop slowly in the course of time, long after sexual maturity has been provided by nature.

Erwin never explained what he believed would be the ideal qualities of a "married wife," but unswerving devotion to himself would have been one prerequisite. Ithi was kind and loving and her spirit was sensuous rather than sensual. She was unique among Erwin's women in the pervasive gentleness of her character. It was inevitable that she would eventually be hurt by a lover so worldly and self-centred, so much her opposite in temperament.

Before they became lovers, Erwin thought that he might be prepared to quit Anny to marry Ithi, but afterwards he was almost horrified by such a thought, though he loved Ithi as dearly as ever, and felt only friendship and pity towards Anny. He thought that "carelessness towards the welfare and wishes of the mate is now a standing feature in [Ithi] – she says herself that she is selfish and she is. What she is longing for in a marital union is her joy. This is an excellent thing if the union comprises nothing but the sex relationship. Sex relations in all their full splendor, and with all the beauty of the world that follows from them."

Even with due allowance for the male supremacist culture of Erwin and his friends, this was hardly an altruistic analysis of their love affair. Ithi proved to be a constant mistress and during four years managed to meet Erwin whenever possible.

Her most dramatic appearance was at a ski lodge in the Alps, where Erwin and Anny were staying. She had just fallen down a crevasse and was battered, bandaged, and bloodstained, when she burst in and threw herself weeping into Erwin's arms. One of their best times together was in the summer of 1932 in Berlin. Ithi came to stay at the Schrödinger house. Anny was conveniently away, and only the maid was there. They went walking, swimming, and rowing together. She met Planck and Einstein. Their carefree happiness ended when Ithi became pregnant. Erwin tried to persuade her to have the child; he said he would take care of it, but he did not offer to divorce Anny. By now, he was thinking more of Hilde March, and already falling out of love with Ithi, and with him even love and the prospect of a family did not suffice for marriage. His continuing marriage to Anny provided a secure defense against importunate mistresses. She was willing to tolerate his every *Seitensprung* (extramarital affair) and acted as an insurance against complications whenever he wished to end one. Even Erwin's desire for a son was not sufficient to cause a break with Anny. In desperation, Ithi arranged for an abortion. Although it was performed by a qualified doctor, it seems likely that damage to the uterus occurred; Ithi later married an Englishman, suffered a series of miscarriages, and was never able to bring a pregnancy to term. When she left Berlin in 1932, she was broken hearted, yet she never lost her loving regard for Erwin.

Some months later, Erwin recorded a strange dream. He and Ithi were going to bed, but there was another person present in the room who kept calling his mother by her first name, always "Georgy, Georgy" but the name was always strangely slurred so that it sounded like "Zhorgie, Zhorgie." This may have been a voice calling his mother to punish a naughty boy, or it may have been the spirit of his unborn son. Erwin and Ithi met briefly once again in London in 1934, but their old love was not revived.

A TREMBLING ELECTRON

Once Schrödinger was settled in Berlin, he turned his attention again to the outstanding problems of quantum mechanics, especially those relating to the reconciliation of quantum theory and relativity theory. In 1928, a major advance had been made by Paul Dirac, who discovered a quantum mechanical equation for the electron that was consistent with special relativity.

Dirac's approach to theoretical physics was quite different from that of Schrödinger; he was not interested in the construction of *anschauliche* physical models, but was content to let mathematics be his guide, being confident that if the mathematical analysis was reasonable the physical picture would in most instances emerge eventually. Thus his approach to the problem of the electron was highly original, and it is difficult to see how anyone else would have thought of it for a long time. In addition to the variables x, y, z, and t, the Dirac equation for the electron included new variables that turned out to describe the electron spin. The equation also allowed solutions for both positive and negative values of energy. At first Dirac thought that this was a defect, but soon afterwards came the discovery of the positive electron (positron) by Carl Anderson at Caltech, and the apparent defect became the greatest asset of the new theory.

For three years, from 1930 through 1932, Schrödinger concentrated on the relativistic theory of the electron, first on extensions of the Dirac theory and then on an attempt to combine quantum mechanics with general relativity theory. By now, however, a host of brilliant younger physicists had entered the field, and they were destined to make the major breakthroughs, creating new theoretical structures in quantum electrodynamics and quantum field theory. Great advances were also being made in the applications of the Schrödinger equation to physical chemistry and solid-state physics, but these did not require radically new theory and were of little interest to Schrödinger.

In one of his Berlin papers, Schrödinger described a curious property of the Dirac electron. It was found to undergo a rapidly oscillating *Zitterbewegung* (trembling motion) which is superimposed on the motion of the center of charge of the wave packet representing the free electron. He found that the amplitude of the *Zitterbewegung*,

$$\xi \approx h/4\pi mc \approx 10^{-11} \text{ cm;}$$

h/mc is the well-known Compton wavelength below which one cannot, in accord with the Heisenberg uncertainty relation, reduce the size of the wave packet without exceeding a momentum mc. There is also an angular momentum associated with the *Zitterbewegung*, so that the electron can be imagined to move through space while describing a tight spiral. A criticism of the model of the trembling electron has been that in reality, an electron can never exist in an isolated state in a field-free space.

DECLINE AND FALL OF THE REPUBLIC

Early in 1929, the German post-war economic recovery began to falter. It was a bitterly cold winter in Berlin and schools had to be closed for a week in February. Unemployment rose to 450,000 in the city and on May Day there was a huge demonstration at Alexanderplatz; the police used tanks to disperse the unemployed, thirty demonstrators were killed and there were many arrests. The Wall Street panic on Black Friday, October 25, presaged a worldwide economic collapse. Early in the new year 1930, there were bloody street battles in Berlin between Communists and Nazis, which neither the city nor the national government seemed able or willing to prevent. The Nazi army (*Sicherungsabteilung*, *SA*) was more disciplined and usually gained the upper hand in battles for control of the German cities. On March 1, the *SA* leader Horst Wessel was killed in a communist attack on his house and the song in honor of this "martyr of the battle for Berlin" became a rousing Nazi anthem. By 1931, Hitler's private army of storm troopers numbered 100,000.

The republican parties feared communists more than they did Nazis and there was a growing feeling that the defense of any capitalist system might require Nazi methods. In the national elections of September, the Nazis emerged as the second largest party, after the socialists, gaining 18.3 percent of votes nationally and 12.8 percent in Berlin.

As the depression deepened, 25 percent of the population of

Berlin was on some kind of relief, and the city finances were tottering. Less than 5 percent of university students came from working-class families, but there was nevertheless considerable privation and even near starvation among the students. There were riots on the university grounds and sometimes it was necessary to suspend classes. The students agitated especially for a quota limitation on enrollment of Jews. Viktor Weisskopf had an office from which he could survey the student fighting in the courtyard, and sometimes he would bring the victims of Nazi assaults into the building and help to clean their bloody heads. The police were not allowed to enter the university grounds so that no help could be expected from them.

The university teachers, in their pleasant suburban homes, were considerably insulated from the effects of disorder in the central city, and also from effects of the economic depression, since as prices fell their government salaries became more valuable. They were not usually political extremists; more than half of them supported republican parties, but many adhered to the German nationalists who were readily converted to the so-called "national socialism."

In 1931, an old girl friend of Anny's, Hansi Bauer, came to Berlin to pursue her art studies at the Charlottenburg Academy. Hansi joined the Schrödingers on skiing excursions to the Riesengebirge, south-west of Berlin, and later took driving lessons with Anny. She was a raven-haired beauty, both rich and talented, and she quickly made many friends in avant-garde circles, enjoying in particular a close relationship with Arthur Köstler, who had come from Vienna by way of Paris to seek his fortune. Convinced that the socialists were "doddering and spineless," he had recently joined the KPD but was working as science editor for the conservative newspaper, *Vossische Zeitung*. Hansi brought Arthur to a party at the Schrödingers, and Erwin at first disliked him intensely: "Why did you ever introduce me to that *Scheiss*?" Later, however, Erwin began to appreciate his interest in science, and he used to send him occasional material for the Ullmann papers.

Shrovetide (*Faschingszeit*), the period just before Lent, was an occasion for carnivals and sometimes quite wild parties. A memorable fancy-dress ball was organized at Harnack-Haus in Dahlem, the residential center of the Kaiser Wilhelm Gesellschaft. Erwin and Anny appeared as Amenophis and Nofretete. The next year the *Fasching* party was held at the Schrödinger house, which was turned into the $\Psi\Psi*$ Hotel, with Erwin as innkeeper. Max Delbrück collected guests at the station in the uniform of a bellboy with a cap emblazoned with $\Psi\Psi*$. Erika Cremer, then a young student with Otto Hahn and Lise Meitner, came as a "dairymaid from the Resonance Mountains" with a pitchfork shaped like a Ψ. The Plancks, Polanyi, the Defants, Grotrian, and many young people were there. It was to be the last of their happy *Fasching* balls.

Although their personal living standards were scarcely affected by the depression, the leading physicists did suffer from severe cutbacks in their research funds, so that there was considerable hardship among the younger untenured staff members. Hendrik Casimir, who was visiting Berlin from Leiden in the summer of 1932, recalls attending with Lise Meitner a lecture by Schrödinger at which an admission fee was charged to raise funds for student aid. Schrödinger tried to show that the tendencies of modern science reflected those in contemporary society. Casimir remarked to Meitner that at least it was a witty talk, but she shrugged and replied *"Ganz witzig,"* her tone implying that the subject was really too serious for humor. At that time Schrödinger made no secret of his dislike for the Nazis and their fascist allies, but he never sought in any way to oppose them actively or to join organizations dedicated to such opposition.

The summer of 1932 marked the beginning of the end for the Weimar Republic. As soon as Papen became prime minister, he lifted the ban on the storm troopers, and in the next six weeks there were almost daily street fights, with 72 killed and 500 severely wounded in Prussia. Student rioting forced the closure of the university. The state government was overthrown by a right-wing putsch, and in the national elections on July 31, the Nazis became the largest party in the Reichstag, but with still far short of a majority. In November, there was another election in which Nazi strength fell markedly from

37 to 33 percent nationally. They had passed the peak of their electoral popularity, but Papen conspired with the Army generals, the Junkers, and some leading industrialists to force the now senile Hindenburg to call Hitler to power. On January 30, 1933, Adolf Hitler was appointed Chancellor of Germany. The transfer of power was accomplished with an observance of legal forms, and the Germans quietly accepted their new master. Most German scientists considered the public expression of any political allegiance to be inconsistent with the dignity of their profession. Einstein was exceptional in his open avowal of socialism and pacifism, as were Lenard and Stark in their vociferous devotion to Hitler. The German universities, however, both staff and students, had been the strongest supporters of anti-semitism, and the Weimar Republic had few friends among the professors.

The hierarchy of professors was not only anti-semitic, it was also anti-socialist, and in Prussia, anti-catholic. These prejudices played an important role in the selection of professors. When Göttingen appointed several distinguished Jewish mathematicians and physicists (Courant, Born, Franck) it was derisively called "the Jewish university." It should be remembered, however, that academic anti-semitism was not a peculiarly German institution. During the 1930s prestigious American universities such as Harvard strove to limit appointments of Jewish professors and admission of Jewish students.

SCIENTISTS AND NAZIS

Despite this political violence and economic deterioration, Schrödinger carried on with his teaching, but he accomplished little research during the dismal academic year of 1932–33. On February 8, he gave a popular public lecture at the Prussian Academy, "Why are Atoms so Small?" Small in relation to what, he asked? His answer was that they are small in relation to the sizes of highly organized living structures that have complex functions, such as cogitating about atoms. Thus the question can be rephrased as: "why are there so many atoms in living organisms?" The answer is that the laws of physics are statistical laws, and they become precise only for systems containing a very large number of particles. If living organisms contained many fewer atoms, they would be subject to random fluctuations that would make any determinate behavior impossible. In view of this general rule, however, the permanence of the genetic elements in living cells after billions of cell divisions is truly remarkable. The explanation must be that the genes are composed of complex molecules in which the atoms are maintained in unchanging structures as a result of the laws of quantum theory. This lecture contained in outline many of the ideas that he was to present ten years later in his Dublin lectures and book *What is Life?*

Many Germans thought that having achieved power, Hitler would moderate some of the more extreme Nazi policies, but they were soon to be disillusioned. On February 27, the Reichstag Building was set afire, probably by the storm troopers (*SA*) as part of the campaign for new elections on March 6, in which, after excluding the Communists, the Nazis gained a majority. On March 24, the new Reichstag passed the "Enabling Act" which voted all power to Hitler and thus, still with some pretense of legality, Germany became a dictatorship.

March 31 was declared to be a day of "National Boycott" of the Jews. Mobs of hoodlums led by storm troopers with swastika armbands roamed through Berlin and other cities preventing anyone from entering shops owned by Jews, and beating up anyone they took to be Jewish. The police stood by laughing and joking. Anny has related that Erwin happened to be downtown in front of Wertheim's, the largest Jewish department store. He was infuriated by the scene and went up to one of the storm troopers and berated him. They turned to attack him, and he would have been seriously mauled had not one of the young physicists been there, wearing Nazi insignia. This was Friedrich Möglich, who after the war became a professor at the Humboldt University in East Berlin. Möglich was able to get Schrödinger away safely.

Einstein was in the United States when Hitler came to power; he declared that he would never return to a Germany ruled by the Nazis. They seized his property and offered a reward for his capture. On the day his ship docked, he wrote to resign his membership in the Prussian Academy. His resignation was accepted, but Bernhard Rust, the Nazi minister responsible for the Academy, demanded that the Academicians formally denounce Einstein. On April 1, without consulting any of the physicists, the Academy published a statement on Einstein, essentially saying that they were glad to get rid of him. Max Laue then summoned an extraordinary general meeting of the Academy and tried to have this statement withdrawn. Only fourteen members attended, and only two supported Laue. Schrödinger did not attend this meeting and took no part in the commotion surrounding the Einstein resignation; in fact, he ceased going to the Academy at about this time, perhaps as a tacit protest. Later the Nazis forced the Academy to expunge the names of Einstein and Schrödinger from its membership records. Even Franck, Born, and other Jewish members did not share this distinction.

On April 7, the Reichstag passed the "Law for Restoration of the Career Civil Service," the purpose of which was to eliminate Jews and socialists from all government positions, including teaching staffs of universities and technical institutes. During the first year, 1,700 faculty members were dismissed. Several efforts were made to persuade Max Planck to make a public protest against the Nazi dismissals of professors. Otto Hahn asked him to support a statement by a dozen eminent physicists, but Planck persuaded him that it would be useless.

There is no known instance in which a professor of physics or chemistry without any Jewish family ever made an open protest against Nazi activities. Among the German intellectual elite, the scientists were conspicuously unanimous in this respect, since a few scholars in other fields did protest. It is true that after 1934 open opposition would have been dangerous, with the establishment of the dreaded concentration camps and revival of what Goering called "the good old German custom of the axe." In the early years of Nazi power, however, opposition was not yet suicidal, and it was during 1933 and 1934 that the scientific establishment, led by Max Planck and Walther Nernst, washed its hands of the growing terror and concentrated on defending its own special privileges. On the twenty-fifth anniversary of the Kaiser Wilhelm Gesellschaft, in 1936, Planck was able to send a telegram to Hitler thanking him for his "benevolent protection of German science."

The only notable German scientist who was conspicuous in his disapproval of the Nazis was Max Laue, and even his actions were taken within the physics establishment and not in open criticism of the regime. Laue tried without success to persuade Heisenberg to help resist the worst Nazi excesses, and called his attention to the biblical injunction against casting pearls before swine.

This unbroken record of collaboration does not mean that every non-jewish scientist was a committed Nazi. Many were merely opportunists who welcomed the chance to advance their careers by taking the positions of dismissed Jews, while the more eminent were either mildly anti-Nazi like Planck and Hahn, or moderately pro-Nazi like Heisenberg. On June 2, 1933, Heisenberg wrote to Max Born urging him not to leave Germany but to take advantage of certain provisions that might exempt him from the general "cleansing" of the civil service.

Since only the very least are affected by the law – you and Franck certainly not, nor Courant – the political revolution could take place without any damage to Göttingen physics . . . In spite of [some dismissals], I know that among those in charge in the new political situation, there are men for whose sake it is worth sticking it out. Certainly in the course of time the splendid things will separate from the hateful.

The "very least" referred to associate professors, *Dozents*, and others, who had not achieved international eminence with Nobel prizes and the like. They indeed had the most difficult time, since foreign universities were not interested in making places for them, and often foreign governments would not even accept them as refugees. Born rejected this advice from Heisenberg with polite contempt.

"THE PROF" TO THE RESCUE

Scientists all over the world anxiously watched these events in Germany but only one of them resolved immediately to do

something. This was the eccentric professor of physics at
Oxford University, Frederick Alexander Lindemann, known
to his closest friends as "Peach" but to everyone else as "the
Prof." His mother was an American and his father an Alsatian,
who had made a great fortune in the construction business.
Lindemann was a bachelor, a teetotaler, a vegetarian, and
played an excellent game of tennis. He had studied with Nernst
and Planck in Berlin, showing early promise in research. In
1919 he was appointed professor of physics at the Clarendon
Laboratory in Oxford, as successor to Clifton, who had served
forty years as a resolute opponent of all research. "The Prof"
himself did little research, but he was good at raising money
and finding talented young staff.

Even his friends did not like "the Prof," owing to his malicious and sarcastic remarks; he could not mention the name of
a woman without suggesting some intrigue. He acted as if he
believed himself to be gifted with a most powerful intellect, and
did not hide his scorn for those less well endowed. Actually he
had an inferiority complex: when he sent his book on quantum
theory to Max Born, he suggested that it would be useful only
as toilet paper. Among the rich, however, he was capable of a
certain charm. Now this may seem a most unlikely man to have
set about a single-handed rescue of Jewish physicists from the
Nazis. He was not himself Jewish, even mildly anti-semitic at
times, and at first his idea was simply to get one or two
outstanding theoretical physicists for Oxford University.

Lindemann soon realized that more than one or two
scientists would require help and he discussed the problem
with the Academic Assistance Council, which had been set up
in London under the chairmanship of Ernest Rutherford. The
council had limited funds, about £13,000 raised by public
subscription. The economy was in the depths of worldwide
depression; many young British scientists could not find jobs of
any kind. The idea of importing a considerable number of
scientists from Germany was unacceptable to the British Ministry of Labour. Now Lindemann's friends in high places
became helpful, and he approached Harry McGowan, chairman of Imperial Chemical Industries (ICI) for financial
support. The positions would be additional to any existent

ones, and thus no British scientists would be displaced. Only scientists of established reputation would be offered places. With these understandings, the Ministry of Labour withdrew its opposition.

"The Prof" had already been to Germany to survey the situation, arriving in Berlin on about April 16. He interviewed many physicists and several chemists, and drew up a tentative list of those who might be forced to leave Germany. He came to afternoon tea at Schrödinger's home, and Erwin expressed freely his disgust with the Nazi policies. Lindemann talked with him about his assistant Fritz London, who was a *Privat Dozent* at the University. It was not yet certain that the racial laws would be extended to such positions, and London was still in some doubt as to whether he should leave. "The Prof" replied that London had asked for time to think it over. "That I cannot understand," said Schrödinger, and added on the spur of the moment, "Offer it to me, if he does not go, I'll take the position."

"The Prof" was astonished by this unexpected news, since he knew that Schrödinger was not Jewish and the Nazis had nothing against him. He was aware that Erwin hated the Nazis, but he had not realized that he was willing to give up his prestigious position in Berlin to face the uncertainties of emigration to England. They discussed the possibilities in some detail and Lindemann promised to try to arrange a suitable position at Oxford. Schrödinger then asked whether a temporary fellowship could also be obtained for his friend Arthur March, an associate professor at Innsbruck. He said that they would like to work together on a book. Lindemann promised to approach ICI on the basis that Schrödinger would need an assistant at Oxford.

At this time Erwin, much in love with Hilde March, was reading Thornton Wilder's *The Bridge of San Luis Rey*, and he confided to his *Ephemeridae* the following quotation:

Now he discovered that secret from which one never quite recovers, that even in the most perfect love one person loves less profoundly than the other. There may be two equally good, equally gifted, equally beautiful, but there may never be two that love one another equally well.

On May 6, the Nazis announced that the racial law would apply to all instructors in institutions of higher learning, and Fritz London was dismissed. He would join Schrödinger at Oxford. Max Born and his family had already left Göttingen and were staying in the Italian Tirol near Bolzano. He wrote to Anny to invite the Schrödingers to visit them there. Northern Italy was to be a sort of assembly place for many of the refugee scientists during the summer of 1933.

The Schrödingers arranged to sell their house and to send their furnishings to England. They sent two large trunks to Switzerland. At this time it was still possible for those leaving Germany to take with them household goods, and even personal jewelry, although the export of money in excess of 1,000 marks was prohibited. Later the Nazis confiscated all the property of anyone who managed to escape.

Anny Schrödinger was an enthusiastic motorist but an inexperienced driver. About the middle of May, she saw a bright new grey cabriolet in a BMW show window in Berlin and fell in love with it. The Schrödingers bought the car, and made plans to travel to the South Tirol.

In Oxford, "the Prof" surveyed the possibilities of a college fellowship for Schrödinger and found that Magdalen College might be able to elect him. On July 21, George Gordon, who was Professor of Poetry and President of Magdalen, wrote to "the Prof" to inform him that the election of Schrödinger had been set down for the next college meeting, on October 3. The post would be like that of a senior fellowship and the holder would be a member of the college governing board. Magdalen is one of the oldest, most famous, and most beautiful of the Oxford colleges. It was founded in 1458 by William Waynflete, Bishop of Winchester. The great tower, completed in 1509, has become a symbol for Oxford itself.

TIROLEAN ADVENTURES

Schrödinger did not formally resign his professorship in Berlin, but he wrote to the Minister of Education to request a "study leave." This letter was not answered at the time. He also sent a postcard to the porter in charge of notices in the physics department to inform him that his lectures would not be given in the fall semester. On September 1, payment of his salary was stopped. His manner of leaving his important Berlin position was like a slap in the face to the regime; even though he made no overt political statement, the Nazis recognized him as an enemy. Yet whenever he was asked why he had left, he gave a standard answer: "I could not endure being bothered by politics."

Schrödinger was exceptional – very few non-jewish professors refused to knuckle under to the Nazis. In the Autumn of 1933, 960 professors published a vow to support Hitler. They were led by Heidegger, the existentialist philosopher. Heidegger was appointed as Rector at Freiburg University and announced that "The much praised academic freedom will be rooted out of the German university."

Erwin had been ardently pursuing Hilde March for some months. He had even offered to divorce Anny, but Hilde did not want to leave Arthur. In May, Erwin wrote in his diary (in French, which must have seemed the appropriate language for amorous intrigue): "Now for the first time I must make a little plot against her. To make her sleep with me. To hold her embraced – even if it is only for one night. It has never happened that a woman has slept with me and did not wish, in consequence, to live with me for all her life. I swear in the name of the good God that it will be the same thing with her." Hilde spent most of the month of June in Berlin. Erwin gave her his diary to read at leisure. This must have convinced her that he really loved and needed her, but she left to rejoin Arthur in Brixen with his "little plot" still in abeyance. On July 6, however, he wrote, "I think indeed she is mine."

About a week later, at the end of the summer semester at the university, Schrödinger was in a somber mood as the day of his final departure from Berlin approached. He was almost forty-six years old, and he reflected "might it not be the case that I have already learnt enough of *this* world. And that I am prepared . . ." Anny, however, was in good spirits as they left Berlin in the little BMW.

They proceeded to Zurich where Anny met Wolfgang Pauli on July 21. She told him that they intended to visit the Borns at Selva Gardena (near Bolzano). Hermann Weyl had resigned his professorship at Göttingen, and also planned to be there. Pauli wrote to Heisenberg to try to persuade him to meet them all in Italy for some mountain climbing. It seems that he also had in mind the possibility of persuading him to leave Germany, but Heisenberg was not to be drawn away. Pauli discussed with Schrödinger the possibility that he might return to Berlin to try to oppose the Nazis. Erwin said only "Ich habe die Nase voll," "I've had a nosefull – I want to get out." Schrödinger was careful never to make a public denunciation of the Nazi regime. He maintained a formally correct relation to the University of Berlin, and later even received a form letter signed by Adolf Hitler thanking him for his services. Never theless he was entered in the Nazi records as "politically unreliable."

By early August, the Schrödingers were installed in Brixen (Bressanone), amid the beauties of their much loved Tirol. The Marches were also there, while the Borns were with their children at Selva, nestled in mountains surrounding the Val Gardena, only about forty kilometers away. Anny went to visit them with her lover Peter Weyl.

Hilde's resistance was at an end, and before long, she and Erwin took off together on a bicycling tour of some of the nearby beauty spots. It is difficult to say to what extent their love affair was founded on mutual passion and to what extent it was based on the more reasoned estimation that they would make highly suitable parents. In any event, when the loving couple returned from their bicycle tour, Hilde was pregnant. Hilde and Arthur had been married for four years and had no children, while Erwin and Anny had been married for thirteen years without children. "I like Anny as a friend, but I detest her sexually," confided Erwin to his diary. Anny's love affair with Weyl was also childless.

This was apparently the only instance in which Erwin had a love affair with the wife of a close friend and professional colleague. Arthur was certainly aware of the situation, but perhaps his admiration for Erwin was so great that he considered it an honor to share a wife with him. When they were all together later that summer at Lake Garda, however, Arthur's discomfiture was revealed by attempts at lighthearted humor which he said were as necessary for his equilibrium as a balancing rod for a tightrope walker, but which tended to make Erwin rather nervous. Anny had long since given up any objections to Erwin's love affairs, especially when Weyl was nearby. Conventional standards of sexual morality were irrelevant. Love and friendship were important and they could co-exist in many permutations without engendering overt jealousy. Tensions inevitably occurred, but they were repressed.

On August 12, Schrödinger wrote to Einstein from Solda, a mountain resort where he was staying with Hilde. Strangely, this letter uses the formal "Sie" instead of his usual intimate "Du," perhaps because he was not sure how Einstein felt about him after the events in Berlin.

After the long time since I last saw you in beautiful Caputh before your American journey, I need to give you at least a sign of life. I hope and believe that I need not fear that the dear and beautiful friendship which you have always bestowed upon me has become dimmed in your thoughts, although unfortunately you indeed have had official cause enough. Spare me from speaking of these things, and let me hope that you know me well enough to make that speaking superfluous . . . I very much hope to see you in Brussels in October . . . Unfortunately (like most of us) I have not had enough nervous peace in recent months to work seriously at anything.

Hansi Bauer, who was on her honeymoon in the Tirol, also visited Lake Garda. She was already somewhat disappointed in her bridegroom, and Erwin advised her to put love above anything else. He met her once in the grocery shop and a "spark of enlightenment" arced between them, which was destined to kindle later a more enduring flame.

As Autumn came to the Tirol, there were snowfalls on the mountains and chilly nights in the valleys. The long lazy summer was over and the refugees began to prepare for departures to their temporary positions, their uncertain futures. Anny and Erwin loaded the BMW and set out via the St. Gotthard Pass for a meeting of the French Society for Physical Chemistry in Paris, where they arrived in the third week of October. Anny bravely drove in the Paris traffic, where Erwin did not dare to take the wheel at all.

The Nazi control of the press had already become so effective that Schrödinger's departure from Berlin was noted in only one German newspaper, the Berlin *Deutsche Zeitung* of October 24, 1933. Under the heading "Loss for German Science," it reported:

The German world of learning has suffered a severe loss: Professor Erwin Schrödinger, the successor of Planck as *Ordinarius* in Theoretical Physics at the University of Berlin, has received a call to the University of Oxford. Schrödinger had already scheduled lectures for the winter in Berlin; hence the call to the English university, which will entail no teaching duties, apparently occurred not without his agreement, and must lead to the conclusion that the distinguished scholar, who has worked in Berlin since 1927, is leaving Germany. This is all the more to be regretted, as only a short time ago Professor Hermann Weyl, *Ordinarius* for Mathematics at the University of Göttingen, also in the field of theoretical physics, accepted a call to the American university of Princeton.

CHAPTER 8

Exile in Oxford

President Gordon had hoped that Schrödinger could arrive in Oxford by October 21, so that he could matriculate and receive the M.A. degree that had been arranged for him (a requirement for the fellowship), but the seventh Solvay Conference was to be held in Brussels, from October 22 to 29, and Schrödinger was obliged to attend. Erwin and Anny stayed at the elegant Hotel Gallia et Britannique, and as usual he arrived looking more like a Tirolean mountaineer than a distinguished scientist.

ARRIVAL IN OXFORD

The Schrödingers and Hilde March arrived in Oxford on November 4. Early in 1934 they moved into a house with a large garden, at 24 Northmoor Road in one of Oxford's best residential areas. The lease on this house (£150 per annum) was taken by ICI and Schrödinger reimbursed the company on a monthly basis. It was just around the corner from the house of Franz and Charlotte Simon. The Marches rented an attractive smaller house, not far away.

Schrödinger found at Magdalen a letter from Luther Eisenhart, Dean of the Graduate School of Princeton University, inviting him to lecture for one, two, or three months at $1,000 a month plus $500 for travel expenses.

Schrödinger's formal welcome as a fellow of Magdalen College was accompanied by some exciting news. According to Armin Hermann:

12. Schrödinger with Lindemann at Oxford
(photograph by Charlotte Simon)

At the conclusion of the ceremony, there was the usual "shake hands" with the other fellows and the almost ritually spoken words "I wish you joy, I wish you joy," and the fellows took their places at the high table for the festive dinner. But at exactly this time, *The Times* of London called Schrödinger's hotel to inform him that the Nobel prize in physics for 1933 had been awarded to Paul Adrien Maurice Dirac and Erwin Schrödinger. The prize for 1932 went to Werner Heisenberg. When they could not reach Schrödinger at the hotel, they called the college.

As Erwin recalled later in a letter to Max Born.

On 9th November, 1933, dear George Gordon, the President of Magdalen College, called me to his office to tell me that *The Times* had said I would be among that year's prize winners. And in his chevalieresque and witty manner, he added, "I think you may believe it. *The Times* do not say such a thing unless they really know. As for me, I was truly astonished, for I thought you already had the prize."

Congratulations poured in from all over the world. Lindemann wrote immediately to ICI "I was amazed to see this a.m.

that Schrödinger has been awarded the Nobel prize . . . He should be given the maximum salary possible . . . £1,000." Owen Richardson, who had received the prize five years previously, wrote from King's College, London, with practical advice about the procedures in Stockholm. He warned Erwin that they would put him up in an expensive suite at the most expensive hotel, the bill for which would be presented just as he was leaving. He would then have a cheque for the prize, but it would be difficult to cash. Richardson kindly offered to lend him some money for the trip if he was short of funds (as he well might have been because of the German restrictions).

NOBEL PRIZE

On December 10, 1896, Alfred Nobel at the age of sixty-three died of a stroke at his villa in San Remo. After accumulating vast wealth through his inventions of dynamite and smokeless powder, he had become a merchant of death who hated war, and a lonely prey to sadistic fantasies who yearned to do something for humanity. The bulk of his fortune of thirty million crowns (8×10^6 at the then current exchange rate) was left to establish annual prizes in physics, chemistry, physiology or medicine, literature, and peace. Each prize was worth about 200,000 crowns ($54,000). This was thirty times the annual salary of a professor and two-hundred times that of a skilled worker, so that the Nobel prizes were the richest ever known; by placing a high monetary value on intellectual discoveries, they greatly enhanced the respect of middle-class citizens for such achievements.

Selection protocols are similar for the different prizes. In the case of physics, nominations are solicited from a select group of scientists, consisting of former prize winners, other noted scientists chosen on an *ad hoc* basis, members of the Swedish Academy of Sciences, professors of physics at the Swedish universities, at Copenhagen, Helsinki and Christiana (Oslo), and at six other universities which vary from year to year. The nominations are considered by a committee of five: the president of the Nobel Institute for Physics and four members elected by the physics section (*Klass*) of the Academy. Committee members are elected for a term of four years, but in most cases appear to have been re-elected for life. The committee report on the nominations, usually with a recommendation for the prize, is considered by a meeting of the *Klass*, which transmits its nomination to a plenary meeting of the Academy. On December 10, the anniversary of the death of the founder, the prizes are awarded in formal ceremonies in Stockholm and Oslo (for the peace prize). The prize winner is expected to give a public lecture on his work.

The will of Alfred Nobel had specified that the prize should be given to the person whose "discovery or invention" shall have "conferred the greatest benefit on mankind." Over the years, this provision gave rise to considerable controversy, since the members of the Academy were almost all engaged in research in basic or "pure" science, whereas Nobel had been an inventor concerned with practical applications. The problem was solved by a gradual acceptance of the theory that pure research leads to an expansion of man's knowledge of the universe and therefore must ultimately confer "the greatest benefit on mankind." In the case of Fritz Haber, who received the chemistry prize in 1919, his development of poison-gas warfare did not delay his award.

The first physics prize was given in 1901 to Wilhelm Röntgen for "the extraordinary services he has rendered by the discovery of the rays subsequently named after him." Until 1915, there was a strong emphasis on experimental work; of the twenty prize winners, only Lorentz, who shared the 1902 prize with Zeeman, was a theoretician. The prevalent feeling was that physics is an experimental science, in which significant advances can be expected only from new discoveries in the laboratory. The absence of theoretical eminence among the Swedish scientists led them to underestimate the importance of the revolution in twentieth-century theoretical physics. The task of evaluating nominations of theoreticians devolved mainly upon Carl Oseen of Upsala. He was a specialist in hydrodynamics, who approached new ideas with extreme caution.

The first nomination of Schrödinger for a Nobel prize in

physics was received in 1927 from an unlikely source, David Starr Jordan, chancellor emeritus of Stanford University and a famous expert on American fishes. He had been invited in October to make nominations for the prizes in physics and chemistry, and after consulting David Webster of the Stanford physics department, he nominated A. H. Compton, E. Schrödinger, and P. Debye, in that order, for the physics prize. Jordan was pleased to learn that Compton had just been awarded the 1927 prize, shared with C. T. R. Wilson, inventor of the cloud chamber.

As early as 1927 Oseen, and hence the other members of the Nobel committee, were aware of the "revolutionary" importance of the work of Schrödinger. Yet Oseen at this time was primarily concerned with the way in which theoretical physics might elucidate the relative importance of different experimental "discoveries." He was reluctant to consider a new theory to be itself a "discovery" worthy of a Nobel prize, and this narrow interpretation of "discovery" would become the basis of his opposition to a prize for the quantum theoreticians, which became increasingly obvious in subsequent years, and finally led to a critical situation as the nominations of Schrödinger and Heisenberg became more numerous and more forceful.

In 1928, the committee noted the nominations of the "pioneers of atomic physics," Heisenberg and Schrödinger, but following the advice of Oseen, it was repelled by the highly mathematical character of their work.

Heisenberg has, from the beginning, abandoned any attempt to create a theory possible to understand with physical thought processes and has been satisfied with mathematical processes in the matrix calculations . . . The same holds for Schrödinger's wave mechanics as can easily be seen from the fact that the space in which his waves propagate is not the usual three-dimensional space. Work on the interpretation of the theory has not led to an experimental discovery of basic importance, nor has it clarified the logical basis of the theories.

The number of nominations received by a candidate does not appear to have strongly influenced the selection committee, although eventually, as in the case of Planck, the "nomination pressure" might become too insistent to be disregarded. Most nominations were one-page letters with concise mention of the work concerned, or even a remark that it was too well known to require comment; only occasionally were detailed supporting statements provided.

For 1929, Jean Perrin, from Paris, who had received the prize in 1926, nominated Louis de Broglie, "who was the first to propose that light and matter have an essentially analogous structure (particles guided by trains of waves) and who predicted quantitatively the frequency of the waves necessarily associated with a moving electron." He included an electron-diffraction picture to emphasize the experimental confirmation of de Broglie's idea. This was the kind of support the committee wanted and de Broglie was awarded the prize. Next year the prize went to C. V. Raman. Despite impressive nominations for Schrödinger and Heisenberg, it was decided to reserve the 1931 award till the following year.

For the 1932 prize, Pauli nominated Heisenberg with the remark: "If I place Heisenberg above Schrödinger for the Nobel prize, it is for two reasons, (1) Heisenberg's matrix mechanics preceded Schrödinger's wave mechanics, (2) Heisenberg's creation must be considered as more original, since Schrödinger could rely for the idea to a considerable extent upon de Broglie." In an interview in 1963, Dirac was asked "Where do you place Schrödinger?" His reply was "I'd put him close behind Heisenberg, although in some ways Schrödinger was a greater brain power than Heisenberg, because Heisenberg was helped very much by experimental evidence and Schrödinger just did it all out of his head."

An important nomination of Schrödinger and Heisenberg came from Einstein. "The contributions of the two men are independent of each other and so significant that it would not be appropriate to divide a prize between them. The question of which should recieve the prize is difficult to decide. I personally value Schrödinger's contribution higher, because I have the impression that the concepts created by him will extend further than those of Heisenberg. If I had the decision, I would give Schrödinger the prize first." In a handwritten footnote to this letter, he added "This, however, is only my opinion, which may be wrong."

A Swedish member once remarked with a mixed simile that "sitting on a Nobel committee is like sitting on a quagmire, one doesn't have a firm foundation under one's feet." Until now the committee had been meekly following the advice of Oseen that award of a prize to the quantum mechanicians would be premature. As more nominations were received from many of the most eminent physicists in Europe, it was becoming obvious that quantum mechanics was probably the most important theoretical discovery of the twentieth century. A whole series of major advances that depended on quantum mechanics could hardly be considered while the basic discoveries remained uncrowned with the Nobel laurels. Yet the opposition of Oseen became even more determined, and in 1932 he prepared a report that finally made evident his intransigence.

I believe I should develop somewhat more closely the reasons that have made it impossible for me in previous years to give the prize to either Heisenberg or Schrödinger. The obstacles that have stood in the way relate both to the by-laws of the Nobel Foundation and also to inner difficulties in the theory. According to the will of Nobel, the prize in physics should be awarded to the person who has made the most important discovery or invention, words which used in the ordinary sense mean either an advance in our knowledge of factual reality or, secondly, a useful application of such knowledge. It is apparent to me that neither of these theoreticians has been able to make such a discovery or invention. The condition for an award must then be that their work has led to some discovery or invention of such importance as to deserve a Nobel prize. In my view, I do not believe that such a condition has been satisfied.

Now that Oseen had taken such a negative stand, the committee thought that an additional report from a less rigid viewpoint would be desirable, and they asked E. Hulthen to prepare a "special submission." He recommended that the prize for 1931 be divided between Heisenberg and Schrödinger.

Despite this dissenting report, the influence of Oseen was still too powerful. The committee members were anxious not to allow any open discussion in the meeting of the *Klass*, which

13. The Nobel party at the Stockholm Station, Anny, Mother Dirac, Paul, Werner, Erwin

would have taken the decision out of their hands. After a long and bitter debate, they persuaded even Hulthen to join in a unanimous recommendation that no prize should be given for 1931 and that the 1932 prize should be reserved till the following year. Thus for two years no Nobel prize in physics had been awarded, even though eminent scientists such as Einstein, Bohr, Planck, and de Broglie had testified to the epoch-making importance of quantum mechanics.

In 1933, many new nominations were received. Schrödinger received nine nominations, including those of five Nobel laureates, and Heisenberg received eight. In the event, a key to the impasse in the committee was provided by the suggestion of Lawrence Bragg that the prize be divided between Heisenberg, Schrödinger, and Dirac. This was the first time Dirac had been nominated. His theoretical prediction of a positive electron had been followed by its experimental discovery. The committee, thus, came to a decision:

The Committee is of the opinion that the point in time has now arisen at which the question of the founders of the new atomic theory should be decided . . . A great discovery from recent times, the discovery of the positive electron, has totally changed one of the most difficult objections against the new theory into a support for it . . . Under these circumstances, a sufficient cause is now at hand to propose an award to the founders of the new atomic physics.

The Committee proposes that the Nobel prize for 1932 be given to Professor Heisenberg (Leipzig) for the presentation of quantum mechanics and applications of it, . . . and that the Nobel prize in physics for 1933 be shared between Professor Schrödinger (Berlin) and Professor P. A. M. Dirac (Cambridge) for the discovery of new forms of atomic theory and applications of them.

STOCKHOLM

Erwin and Anny arrived in Stockholm on December 8, two days before the formal presentation of the prizes. This took place on Sunday evening in the grand ballroom of the Concert Hall, which was decorated with an array of Swedish flags and a profusion of flowers. Exactly at 5:00 p.m., King Gustav arrived, accompanied by other members of the royal family. The royal anthem was played by the orchestra and sung by the standing audience. Then the laureates entered, escorted by Swedish prize winners of past years. Each laureate received a diploma, a gold Nobel medal, and a medal from the Academy of Sciences. The obverse of the Nobel medal was a portrait of the founder by Erik Lindberg, and the reverse bore the name of the laureate and an inscription from the *Aeneid*: "*Inventas vitam juvat excoluisse per artes.*" (It is delightful to discover life embellished by the arts.) The Academy medal portrayed Nature as the goddess Isis emerging from clouds and holding in her arms a cornucopia, while the veil which covers her face is lifted by the Genius of Science.

The next event in the festivities was the ceremonial banquet in the Winter Garden of the Grand Hotel. The Prince and Princess Royal and all the princes and princesses were there but not the King and Queen. Heisenberg and Dirac were accompanied by their mothers. After formal toasts to the King and to the memory of Nobel, toasts were offered to the prize winners, who replied at various lengths. Dirac's words were somewhat incongruous. He said that the world was in the

depths of economic misery, but anything to do with numbers
should be capable of theoretical solution, and that the cause of
the great depression was that people much preferred to collect
interest "through all eternity" instead of taking a single
payment for their goods or services, and hence there was a
great shortage of buyers for the world's goods. Anny thought
that his talk was "a communistic propaganda tirade." Heisenberg spoke briefly to thank everyone for the hospitality.

Schrödinger, perhaps with some inspiration from the champagne, gave the most spirited response, full of "enthusiastic
ardor":

Put yourself please in the place of a man who for the first time wakes
up early in the morning with the bright sunshine to set foot in this
wonderful city with its expanses of water, splendid buildings, its
proud castle, its stones and spires, the happy, kindly people, who
accept you with friendship; this city, which is so modern, so different
from any other in the world, so that it has you immediately under its
spell, and you want only to walk about and look and climb here and
there, in order to drink in its picture as fully and completely as
possible. Yes, one can do nothing else, one does not hear the call of
duty, but thinks that somehow or other tomorrow will go well –
people who live in such beauty will surely not be such stern critics.

The next evening there was a smaller dinner given by the
royal family at their palace. There were a hundred guests and
forty servants to wait upon them. The King and the crown
princess were served by a special servant wearing a high hat
with feathers. Last night the menu had been consommé, sole,
chicken, and ice cream, but tonight they dined on caviar,
salmon, pheasant, and venison. After the dinner there was a
reception for 1,200 people at the National Gallery. The Nobel
festivities are the high point of the Stockholm social season, and
the entire city takes part in them.

Schrödinger's formal Nobel lecture was given on December
12, on "The Fundamental Idea of Wave Mechanics." The
lecture was intended for an audience interested in science but
without specialized knowledge of mathematical physics. It was
essentially a more popular account of the way in which he had
developed the analogy between optics and mechanics in the

second of his great Zurich papers of 1926. It was a modest lecture in that it did not give any emphasis to his own contributions to the subject and the revolution they wrought in physics and chemistry.

When Erwin and Anny returned to Oxford, they found a Christmas present addressed to Erwin from Peter Weyl. This was a fine edition of Shakespeare's *Venus and Adonis*, with beautifully erotic illustrations. Peter had written an inscription: "The sea has bounds but deep desire has none."

PRINCETON

At the beginning of the fall term of 1933, the Physics Department at Princeton University was considering how to fill the Thomas D. Jones Professorship of Mathematical Physics. This was one of Princeton's most distinguished professorships. The outstanding possibilities were Heisenberg and Schrödinger. He accepted the invitation to visit Princeton, and embarked for the United States March 8 on the *President Harding*, returning to England on April 13.

While in Princeton, Schrödinger stayed at the Graduate College, a structure in collegiate Gothic style in imitation of an Oxford college, with tower, common room, a large refectory with stained glass windows, and accommodation for students and fellows. The English atmosphere was not recreated perfectly, however, since the plumbing was modern, no alcoholic drinks were allowed in the public rooms, and the servants were not old British characters with lower-class accents, but young and breezy Cypriots with an erratic command of American slang. Since Erwin did not care for the English original, it is doubtful that he was much impressed by the American imitation.

Schrödinger's lectures, as usual, were models of scientific exposition, and, early in April, he was offered the Jones Professorship. He wrote to Lindemann to advise him of the situation:

Other things have come up here, on which we shall have to have *ausführliche* [extensive] talks after my return. There is not much use discussing them in a letter, yet since – in spite of all confidentiality – rumours spread out almost in contradiction with the principle of light, I should like to avoid your hearing them in this way. Yesterday a permanent chair at this university was offered to me. Strange situation! It is only in three weeks from now that my wife will be busy to move us at last into definite and permanent quarters in Oxford. And precisely during these weeks I am forced to take into preliminary consideration a new aspect, under which the state reached in Oxford would not be so very "permanent." I am not frightfully happy at that, yet I am aware that I have to consider things seriously. For after all and in spite of all so-called scientific reputation, I am actually without what a man of my age and métier considers a *Lebensstellung* [permanent position]. And if, e.g., I were drowned on the passage, I am afraid that my wife could neither live upon the German pension nor on the *Schrödingergleichung* [Schrödinger equation]. .

Schrödinger finally decided to decline the Princeton professorship. He gave the reason that the salary and pension were not sufficiently agreeable to him. He might have accepted a position like Einstein's at the Institute for Advanced Study. As he remembered his dire financial situation in post-war Vienna, old age pensions became something of a fetish for him – his worry about future disasters prevented a reasonable view of present opportunities. He still had his Nobel prize safe in a Swedish bank. He was one of these fairly well-to-do people who always feel afraid of poverty.

The reasons given by Schrödinger for declining the prestigious American professorship may not tell the whole story. There have been persistent rumors at Princeton that he discussed with President Hibben his plans for an unusual family arrangement, in which he would live with Anny as an official wife and Hilde as a second unofficial wife, both women sharing the care of his child. He wondered if this extended family would be acceptable to the Princeton community, and he even worried about a New Jersey law against bigamy. He wrote in his journal that it would be sad to go away and "leave the mother and the child." Since he had been served ice-water with his oysters in 1927, Erwin had never overcome his aversion for the American way of life and, after Vienna, Zurich, and Berlin, Princeton must have seemed somewhat as it did to

Einstein, "a quaint ceremonious village of puny demi-gods on stilts."

LIFE IN OXFORD

During term time at Oxford, Schrödinger was scheduled to give one lecture a week, on "Elementary Wave Mechanics." This was held at the Clarendon Laboratory at noon on Saturdays. The lecture hall was icy cold in winter, warmed only by the steamy breaths of the students. Unlike the German system in which all of physics was covered systematically in comprehensive lecture courses, the English universities relied upon a few schematic lectures which were intended to suggest lines of independent study. Often even the few lectures provided were not especially good. Lindemann lectured twice a week on different topics each term, but owing to a slight speech impediment, his lectures were rather unintelligible. Schrödinger thought that these teaching arrangements compared unfavorably with those in Germany, and he often complained to Anny that he was being paid for doing nothing or that he was being treated like a charity case. His lectures, though few, were highly appreciated and were said to be the best physics lectures ever heard in Oxford.

It was late Spring in Oxford when Erwin's first child, a daughter, was born. Hilde March had entered the Maternity Home attached to the Acland Hospital and the baby was born on May 30, 1934. She was christened in the Church of England as Ruth Georgie Erica. The middle name was in memory of Erwin's mother Georgine who was always called Georgie. The birth was registered by Hilde on July 2, and of course Arthur March, professor of physics, was named as the father.

It is said that Hilde suffered a post-partum depression and tended to reject the baby. Anny, however, was very supportive and for several months took over much of its care. Once while watching a nurse change the two-week-old infant, Erwin remarked, "Goodness, what a lot of things you need to treat such a little mite correctly." Matter-of-factly she replied, "Oh,

14. Anny Schrödinger in Oxford

I suppose so, but it takes quite a lot to kill it." He remembered this remark twenty-five years later.

Before the birth of the baby, Erwin and Hilde used to go everywhere together in Oxford, and he made no attempt to conceal their special relationship. He did not regard her as a mistress, but rather as a second wife who happened to be married also to another man. Conventional sexual morality was simply not worth bothering about, so long as those directly concerned accepted the situation. His understanding of personal feelings was often naive. More than one of his friends have even called it "childish."

Erwin loved and appreciated women, but his attitude toward them was essentially that of a male supremacist. Oxford society, on the other hand, was based on a sort of official misogyny. Wives were regarded as unfortunate "female appendages." Alfred J. Ayer who was a lecturer in philosophy at Christ Church College, where Lindemann lived, has described Oxford society at this time in his autobiography:

It was still almost wholly masculine . . . The men had spent most of their time at boarding schools in a homosexual atmosphere. They were ill at ease with women. Such sexual experience as they had was almost always homosexual. The active homosexuals were not a majority of Oxford men, but they were very much in evidence. The tone was set by a number of celebrated "Queens," whose flamboyant appearance was joined to a studied formality of manner. Many of those who paid court to these queens were not radically homosexual, but simply continuing their schoolboy practice of using boys as a substitute for girls . . . Life in the colleges was organized on monastic lines. The good food and drink were for men only. If fellows had wives, they were not admitted. Many of the permanent staff were homosexual.

Erwin regarded a society without women as detestable and barbaric. He complained to Max Born, "These colleges are academies of homosexuality. What queer types of men they produce." He did not enjoy the college dinners. "You never know who your neighbor might be. You talk to him in your natural manner, and then it turns out that he is an archbishop or a general – huh!"

It is not surprising that Erwin found it difficult to adapt to Oxford society. When Lindemann discovered Erwin's liaison with Hilde March, he became furious, with words to the effect that "He said he wanted to bring the husband here as his assistant, but he was really after the wife. We ought to get rid of the bounder." It was deplorable to have one wife at Oxford – to have two was unspeakable. Relations between "the Prof" and Schrödinger became strained. Lindemann was trying to get ICI to renew the grants for refugee scientists beyond the two years originally promised. ICI was reluctant to spend more money, but finally agreed to an extension through 1936. One director said they already had been extremely generous since they had paid not only for the scientists but in some cases for their mistresses.

SPAIN

In the summer of 1934, while Anny went to Switzerland to be with Weyl, Erwin visited Spain, where he had been invited to give a course of lectures at the Universidad Internacional de Verano in Santander, which was directed by the philosopher José Ortega y Gasset, and also to lecture in Madrid. Ortega had recently gained international recognition outside the field of philosophy by his book *La rebelion de las masas*, which expressed in rather flamboyant terms his aristocratic view that democracy posed a growing threat to authentic personal existence.

Schrödinger's lectures were intended to explain the fundamental ideas of wave mechanics to a non-mathematical but somewhat philosophical audience. They are written in an unusually informal style, and one can almost hear the author explaining the deepest questions with a charm and lucidity that no other physicist could have equaled. His lectures on the vectorial representation of the wave functions (in Hilbert space) and on the equivalence of wave and matrix mechanics are masterpieces.

Schrödinger was impressed by the personality and ideas of Ortega who, despite his aristocratic background, was a loyal citizen of the Spanish Republic. Ortega thought that politics and government are inevitably degrading and the sensitive intellectual must avoid such activities and devote himself to "rational and responsible" personal relations like love and friendship. It is doubtful, however, that Schrödinger would have gone so far as to accept the view of Ortega that "philosophy, science, and mathematics are pure exact fantasy, games played according to strict but arbitrary rules, by a minority who seek to escape the tedium, vulgarity, and deadly seriousness of the world of beliefs."

Schrödinger liked Spain so much that he decided to return as soon as possible. The following Spring, he set out with Anny in the little BMW to make a wider tour of the peninsula. They covered a great figure eight, with the midpoint in Madrid, south through Valencia, Gibraltar and Cadiz, north through Salamanca, Altamira, and Roncevalles, altogether about 8,000 kilometers. They were delighted with everything, especially the hot sunshine. Erwin was astonished by the prehistoric paintings in the caves at Altamira, which he thought "were not talented children's drawings, but a free flowering of powerful forms."

In Madrid Schrödinger gave a series of lectures which were received with enthusiasm, so that Blas Cabrera and the other Madrid physicists began to explore the possibility of attracting him to a permanent professorship. While in Madrid he wrote to the authorities in Berlin to request a formal release from his professorship there and the grant of emeritus status. On March 31, his resignation was accepted; on June 20 Hitler sent a formal letter of thanks for his services; in July he was given the status of professor emeritus. These amicable arrangements must have led him to believe that he was not in the bad books of the Nazi regime.

SPOOKY ACTION AT A DISTANCE

In the May 15, 1935 issue of *Physical Review* Einstein launched a subtle new attack on the Copenhagen interpretation of quantum mechanics. The paper, "Can Quantum-Mechanical

Description of Physical Reality be Considered Complete," written with two young co-authors, Boris Podolsky and Nathan Rosen, has come to be called the *EPR paper*. EPR did not question the correctness of quantum mechanics but its completeness. "Whatever the meaning assigned to the term *complete*, the following requirement for a complete theory seems to be a necessary one: *every element of the physical reality must have a counterpart in the physical theory* . . . The second question is thus easily answered, as soon as we are able to decide what are the elements of the physical reality."

EPR then proposed a sufficient condition for an element of physical reality: *if, without in any way disturbing a system, we can predict with certainty* . . . *the value of a physical quantity, then there exists an element of physical reality corresponding to this physical quantity.* This criterion rests upon the assumption that the physical world can be correctly analyzed in terms of distinct and separately existing elements of reality. It has been called the assumption of *local realism*.

The rest of the EPR argument will be presented in a form different from that of the original paper, merely as a simplification. Electrons and other elementary particles have an intrinsic angular momentum or spin, so that in a magnetic field they act as little magnets that can take only either one of two orientations to the field direction, which may be called + or −. It is the experiment of subjecting the particle to a magnetic field that causes it to declare itself as + or −. It is a basic tenet of quantum mechanics that the particle itself does not possess the property of being + or −; before the experiment it has a probability 0.5 of being + and an equal probability 0.5 of being −. If two particles, designated a and b, are brought together so that they interact, they will usually go into a state of lowest energy, called the *singlet state*, in which if one particle is +, the other is −. There is no way to say which is which, and the wave function to express this situation is written as $(1/\sqrt{2})[a(+)b(-) - a(-)b(+)]$.

Now suppose that after their interaction the particles have separated to such a great distance that there is no possibility of any further interaction between them. Because of their inter-action in the past, the particles are an example of an *entangled system*. An experiment is now made to measure the spin orienta-tion of a. If it is found to be a(+), it is absolutely certain that b must be b(−). EPR now apply their criterion of physical reality: since the value b(−) was predicted for the spin of b without in any way disturbing b, it must correspond to an existent element of physical reality. Yet this conclusion contra-dicts a fundamental postulate of quantum mechanics, accord-ing to which the sign of the spin is not an intrinsic property of the particle, but is evoked only by the process of measurement. Therefore, EPR concludes that quantum mechanics must be incomplete. For example there may be "hidden variables" not yet discovered, which determine the spins as intrinsic prop-erties of the particles.

On June 7, Schrödinger dashed off a letter to Einstein: "I was very happy that in the paper just published in *P.R.* you have evidently caught dogmatic q.m. by the coat-tails." He concluded: "My interpretation is that we do not have a q.m. that is consistent with relativity theory, i.e., with a finite transmission speed of all influences. We have only the analogy of the old absolute mechanics . . . The separation process is not at all encompassed by the orthodox scheme." Schrödinger had recognized the essential point of the EPR paradox – the separ-ation process and hence the assumption of local reality.

Einstein wrote: "You are the only person with whom I am actually willing to come to terms. Almost all the other fellows do not look from the facts to the theory but from the theory to the facts; they cannot extricate themselves from a once accepted conceptual net, but only flop about in it in a grotesque way."

Einstein proposed that the Ψ function does not describe the state of an individual system but statistically the state of an ensemble of systems.

Naturally this interpretation of quantum mechanics displays especially clearly that, through its restriction to statistical statements, it has necessarily accepted the feasibility of only an incomplete representation of real states and processes . . . You, however, see something quite different as the reason for the inner difficulties. You see in Ψ the representation of reality and would like to change its connection with the concepts of ordinary mechanics or do away with them altogether. Only in this way could the theory be made to stand on its own two legs.

Einstein did not believe that Schrödinger's approach would overcome the difficulties. He asked him to consider the case of a mass of gunpowder that would probably explode spontaneously in the course of a year. During this time the Ψ function would describe a sort of superposition of exploded and unexploded gunpowder. "There is no interpretation by which such a Ψ function can be considered to be an adequate description of reality." Einstein's gunpowder would soon reappear in the form of Schrödinger's cat.

Leon Rosenfeld was in Copenhagen when the EPR paper arrived. He recalls that

the onslaught came down upon us as a bolt from the blue. Its effect on Bohr was remarkable . . . We were in the midst of important theoretical work . . . A new worry could not come at a less propitious time. Yet, as soon as Bohr had heard my report of Einstein's argument, everything else was abandoned: we had to clear up such a misunderstanding at once. We should reply by taking up the same example and showing the right way to speak about it. In great excitement, Bohr immediately started dictating to me the outline of such a reply. Very soon, however, he became hesitant. "No, this won't do, we must try all over again . . . we must make it quite clear." So it went on for a while, with growing wonder at the unexpected subtlety of the argument . . . "We must sleep on it." The next day he was calmer. He put aside everything else and worked on his reply for six weeks.

This reply was published in *Physical Review* (October 15, 1935). Bohr rejected the criterion for reality as stated by EPR. "There is . . . no question of a mechanical disturbance of the system under investigation during the last critical stage of the measuring procedure. But even at this stage there is essentially a question of *an influence on the very conditions which define the possible types of predictions regarding the future behavior of the system.*" Bohr maintained that the spin of the particle is created only by the process of its measurement, and presumably the same might apply to the explosion of a mass of gunpowder.

SCHRÖDINGER'S CAT

Motivated by the EPR paper, Schrödinger published in 1935 a three-part essay in *Naturwissenschaften* on "The Present Situation in Quantum Mechanics." He said he did not know whether to call it a "report" or a "general confession." It is written in a sardonic style, which suggests that he found the "present situation" to be less than satisfactory. It would be his definitive statement about the theory that he and Heisenberg had created.

He first explained in detail how physics, on the basis of experimental data, creates *models*, which are representations of natural objects idealized or simplified so that mathematical analysis can be applied to them. In quantum mechanics, not all the variables for a model can be simultaneously specified. If one measures exact values for the position co-ordinates, one can determine nothing about the values of the momentum components. This situation is a result of the Heisenberg uncertainty relation, which is derived directly from the fact that the operators for position q and momentum p do not commute. It is, however, possible to measure values of q and p that fall within certain ranges in accord with the uncertainty relation, so that one can speak of the specification variables of the model as being washed out or blurred. Nevertheless the wave function Ψ defines the *state* of the model unequivocally. It constitutes a complete catalog of the probabilities of finding any specified result for a measurement made upon the physical system for which the model was designed.

Schrödinger asks "are the variables really blurred?" He points out that the classical description with its sharp values for the variables can be replaced by the Ψ function as long as the blurring is restricted to atomic dimensions which escape our direct control. But when the uncertainty includes visible and tangible things, the expression "blurring" becomes simply wrong.

One can even construct quite burlesque cases. A cat is shut up in a steel chamber, together with the following diabolical apparatus (which one must keep out of the direct clutches of the cat): in a

Geiger tube there is a tiny mass of radioactive substance, so little that in the course of an hour *perhaps* one atom of it disintegrates, but also with equal probability not even one; if it does happen, the counter responds and through a relay activates a hammer that shatters a little flask of prussic acid. If one has left this entire system to itself for an hour, then one will say to himself that the cat is still living if in that time no atom has disintegrated. The first atomic disintegration would have poisoned it. The Ψ-function of the entire system would express this situation by having the living and the dead cat mixed or smeared out (pardon the expression) in equal parts.

It is typical of such cases that an uncertainty originally restricted to the atomic domain has become transformed into a macroscopic uncertainty, which can then be resolved through direct observation. This inhibits us from accepting in a naive way a "blurred model" as an image of reality . . . There is a difference between a shaky or not sharply focused photograph and a photograph of clouds and fogbanks.

This conclusion has been called *the principle of state distinction*: states of a macroscopic system which could be told apart by a macroscopic observation are distinct from each other whether observed or not.

Only a few commentators on the cat paradox, the most notable being Eugene Wigner and John Neumann, have defended the uncompromising idealist position that the cat is neither alive nor dead until a human observer has looked into the box and recorded the fact in a human consciousness. It might be, of course, that the cat itself has a consciousness adequate to complete the experiment and resolve the prob-abilities by passing from a superposition of two states to a single state.

Even without an animal consciousness, the experiment might be decided as soon as the atomic disintegration activated the Geiger counter. In the state $\psi_A + \psi_B$, the waves ψ_A and ψ_B must represent solutions for the time-dependent Schrödinger equation for the macroscopic system, including the cat. For superposition of states to occur, these wave functions must remain in phase. If their phase coherence is lost, the mixed state is no longer permissible. Owing to interaction with the surroundings, the coherence time for such macroscopic wave functions would be vanishingly small, and it is meaningless to talk about a superposition of states, ψ (live) $+ \psi$ (dead). The cat paradox, however, was useful as an antidote to the view that the wavefunction refers not to a physical model but to human knowledge about an object, and perhaps even more importantly, it served to emphasize the principle of state distinction.

The subsequent history of the EPR paradox would fill several volumes. In 1935, no physicist would have thought it worthwhile actually to perform an EPR-type experiment, because they were all convinced that the result would in no way be able to decide between the ideas of Einstein and those of Bohr. Then, in 1954, John Bell, an Irish physicist, derived an important theorem which shows how an experimental decision can be made between the quantum mechanical predictions about entangled systems and the predictions of theories based on local realism. Some beautiful experiments were made by John Clauser, Alain Aspect, and others, and the results showed that a measurement on one of a pair of particles in an entangled system does evoke a correlated property in the other particle. This result has been found even when the particles are separated by a distance so great that even an influence at the speed of light could not reach the second particle. These results appear to have left three options for theoretical physics:

(1) To deny the validity of inductive logic, so that it would be impossible to predict the results of future experiments from those already done. (2) To deny that the microscopic entities of physics have any objective reality; they would exist only in the mind of the observer. (3) To assert that an influence can be propagated from one part of an entangled system to the other at faster than the speed of light. Such superluminal correlation is equivalent to a denial of Einstein's postulate of local realism.

Most physicians, with varying degrees of reluctance, have opted for the third alternative. It means that the reality behind the appearances of physics is an unbreakable whole. If the physical world is real, it is holistic – it is not merely the sum of separable parts.

Schrödinger never drew such a conclusion from quantum

mechanics. He believed that the problem was caused by the extension of non-relativistic quantum mechanics beyond its legitimate range of application, which should be to distances small enough to permit neglect of the time light takes to travel across the system. Many present-day interpretations of the EPR experiments do not agree that the "spooky action at a distance" can be ascribed to the neglect of special relativity. Rather it is believed to be an ineluctable part of the quantum world. Besides the physical paradox of quantum mechanics, we are faced with a psychological paradox in the mind of its creator. His religion of Vedanta taught the unity of the world, he was to devote the latter part of his life to research on unification of physical field theories, he rejected the duality of wave and particle in favor of a world of waves alone, yet he was unwilling to accept the indivisible nature of the quantum world.

In March, 1936, Schrödinger met Bohr in London, and reported to Einstein:

Recently in London spent a few hours with Niels Bohr, who in his kind, courteous way repeatedly said that he found it "appalling," even found it "high treason" that people like Laue and I, but in particular someone like you, should want to strike a blow against quantum mechanics with the known paradoxical situation, which is so necessarily contained in the way of things, so supported by experiment. It is as if we are trying to force nature to accept our preconceived conception of "reality." He speaks with the deep inner conviction of an extraordinarily intelligent man, so that it is difficult for one to remain unmoved in one's position.

AN UNCERTAIN FUTURE

The ICI grants were due to expire at the end of 1935, and in October the company notified all the grantees that they would not be renewed. A special exception was made for Schrödinger and he was awarded a two-year extension. Franz Simon wrote to Born that one should not blame ICI alone, since the company had always stated that the grants were meant merely to provide a buffer against an emergency situation, and they expected that the refugees would be absorbed into more permanent positions. Unfortunately this had not happened in some cases.

Arthur, Hilde, and Ruth March had returned to Innsbruck, where Hilde made a stay of several months in a sanatorium, until her physical and mental equilibrium was restored. She was an unsophisticated woman, without intellectual pretensions but with a great zest for life, and the stress of living with two men, with their often conflicting demands upon her, had temporarily exhausted her psychological reserves.

Meanwhile, Hansi Bohm had escaped from Berlin with her husband and was living in London. In addition to her drawing, she had now reached a professional standard in photography. She became a frequent visitor to Oxford. Anny had taken a small flat in London, so as to take some special courses, and also to give Erwin more freedom to be with Hilde or other friends. Erwin had always been attracted to Hansi, who was so different from the other women he had known. Since his youthful love for Felicie he had not had a close relationship with a woman whose social class, or at least her financial status, was definitely superior to his own. Hansi combined the cultural background of a wealthy Viennese Jewish family interested in music and art, with the more daring sophistication of avant-garde Berlin before it was crushed by the storm troopers. Although Hansi found the Schrödingers conventionally middle class in their artistic tastes, she could not escape the fascination of Erwin himself, his brilliance of intellect, his insights into feminine psychology, his boyish elfin charm. The spark of recognition in the Tirolean grocery shop now kindled a flame of passion. It was inevitable that they should become lovers, and after Hilde left in 1935, Hansi and Erwin went on a short holiday together to the Channel Islands. He was delighted with her both intellectually and amorously – a combination of satisfactions that he had not previously experienced.

Although his personal life had its compensations, Schrödinger was becoming increasingly dissatisfied with his professional situation. A fellowship at an Oxford college and a grant from ICI hardly equaled the professorship at an impor-

15. Hansi Bauer-Bohm (c. 1933)

tant university which his achievements deserved. He had been offered a professorship at Madrid, and he was considering this seriously when the Spanish civil war erupted on July 18, putting an end to that particular future.

In May, Schrödinger had written to Einstein about his interest in a professorship that would soon become available at the University of Graz. He said that he would be inclined to accept this chair provided certain financial provisions were made, such as freedom to keep his securities in Sweden and exemption from tax on travel allowances from abroad. (His Swedish assets amounted to $12,000.) He may have been politically naive, but he had become financially sophisticated.

It is not that I can't stand it in one place for long. Up till now I've generally been contented wherever I was except in N[azi] Germany. Also it is not that they haven't been very nice and friendly to me here. But nonetheless the feeling grows stronger of having no employment and living on the generosity of others. When I came here I thought I would be able to do something for the teaching, but no value is placed on that here. And if I think more about it, really I must say to myself, I am sitting here waiting for the demise or the complete decrepitude of a very dear old gentleman (Love) and the possibility that they might make me his successor. I don't want to be hypocritical that this hurts my feelings, but it does hurt my self-esteem.

In December, 1935, Schrödinger visited Austria. He stayed in Graz with his old friend Kohlrausch and they discussed the possibility of a professorship there. On December 18 and 20, he lectured in Vienna, and consulted the Ministry of Education about appointments in Vienna and Graz. The Christmas holidays were spent skiing at Obergurgl, and he had an opportunity to be with Hilde and Ruth again. It also seems likely that he made a brief visit with Max Laue to Berlin, for Laue's son Theo, who had then just obtained his driving license, remembers bringing him from the station at about that time. He returned to Oxford via Brussels.

EDINBURGH 1936

Towards the end of 1935 an important professorship of physics became vacant at the University of Edinburgh when Charles

Darwin, grandson of the biologist, resigned from the Tait Chair of Natural Philosophy to become Master of Christ's College, Cambridge. The professorship had been established in 1922 to perpetuate the memory of Peter Guthrie Tait, a Scottish mathematical physicist who, among other noteworthy accomplishments, had worked out the dynamics of flight of the golf ball. In February 1936, a committee on the future of the chair was convened under the chairmanship of Edmund Whittaker, and in May it reported to the University Court that "they had considered a number of names, and were unanimously of the opinion that of those available Dr. Erwin Schrödinger was much the most distinguished, and unless it was considered undesirable to appoint a person of other than British nationality, they recommended that the Chair be offered to him conditionally on the Court being satisfied that he would be permitted to continue residence in this country." The Court approved this report and authorized the Secretary to communicate the conditional offer to Dr. Schrödinger.

Some weeks before this formal offer, Schrödinger arrived in Edinburgh to "spy out the land." He was wearing one of his Alpine costumes as protection against the Scottish weather; Whittaker met him at the railway station and was surprised by the informality of his attire. Darwin showed him around the university and explained the heavy administrative and teaching duties expected of a professor in Scotland. Schrödinger must have expressed considerable interest in the post, since the formal offer was sent to him. He told Hansi that he would go to Edinburgh if she would come with him, but this was a completely impractical suggestion, since she was expecting her first child and was planning to return to Vienna.

Perhaps Erwin would have accepted the Edinburgh post, even without Hansi, had bureaucratic incompetence not intervened. The university Secretary wrote to the Home Office to secure permission for Schrödinger's permanent residence. For some reason "an accidental delay" occurred. As time passed with no decision from London, Schrödinger became less attracted by the offer. The salary of £1,200 was much less than that of the Princeton professorship he had declined the year before;

however, the retirement age was seventy and the pension provisions were adequate. Meanwhile an offer had come from the University of Graz with an honorary professorship at Vienna, through the efforts of his friend Hans Thirring, and he decided to accept this. He told Darwin that the call of the mountains of his native Austria was too strong to resist. Also the Austrian pension provided full salary after retirement. Hilde and Ruth were back in Austria, and Hansi would soon be there, so that, everything considered, the opportunity to return home was irresistible. The Tait Chair was then offered to Max Born, who accepted it gladly.

CHAPTER 9

Graz

Niels Bohr was fond of quoting "an old Danish proverb" to the effect that "prediction is always difficult, especially of the future." Schrödinger later described his decision to return to Austria in 1936 as a miscalculation of the political situation that was "an unprecedented stupidity," but it would not have seemed so at the time. He was not satisfied with the quality of life in either Britain or America, and old friends were urging him to return to his native land. Nevertheless, he knew that in going to Graz, he was stepping into a cesspool of Nazi activity. The University was a center for the Styrian Nazi party, whose leader was Armand Dadieu, professor of physical chemistry. More than half the students were active Nazis and they dominated the campus. The principal Graz newspaper, *Der Tagespost*, followed the Nazi line and the provincial *Heimwehr* was controlled by the party.

THE AUSTRIAN SITUATION

During the fifteen years that Schrödinger had lived abroad, the history of Austria had been marked by political strife between two major forces, the Christian Socialists or "blacks" and the Social Democrats or "reds." The clerical party was traditionally anti-semitic, for example, in 1933, its chairman announced that "the religious German must decisively reject baptism as an 'entrance ticket' for Jews." There were about 190,000 Jews in Austria, with 176,000 in Vienna (9 percent of city population) where they were prominent in learned professions, the press, music, and the arts. As in Germany, the universities were citadels of anti-semitism and it was difficult for even distinguished Jewish scholars to obtain professorships. In 1932, the clericals selected Engelbert Dollfuss as chancellor. This diminutive politician (*Millimetternich*) was motivated by a colossal vanity and a paranoid hatred of the Social Democrats who comprised almost 50 percent of his compatriots. At the urging of Mussolini, he suspended parliament on March 4, 1933, and established a fully fascist one-party state, committed to the fight against two internal enemies, the "reds" and the Austrian Nazis. Britain and France urged him not to oppose Hitler, and his only support came from Italy. In 1934, Mussolini asked Dollfuss to crush the socialists and on February 12, Major Emil Fey led the army and police in an attack on the workers' suburbs in Vienna, in which about 1,000 people were killed, including many women and children. As soon as the socialists were destroyed, the Nazis stepped up their terrorism, and on July 25 they assassinated Dollfuss, but the putsch then failed, as Hitler did not intervene as expected, having been warned off by Mussolini.

Kurt Schuschnigg now became leader of Austria, a position he was to hold for almost four years, until he handed the country over to Hitler. As long as Mussolini was able to support him, his position remained fairly secure, but as the Italian dictator became embroiled in Abyssinia, and as Hitler's military strength grew, Nazi pressure on Austria became more insistent. This was the situation in Austria as Schrödinger was negotiating the conditions for the professorship at Graz. He had no sympathy for the black dictatorship, and its excesses and incompetence confirmed his contempt for politics in general.

The Schrödingers rented a large house in Graz. Hilde and Ruth were in Innsbruck with Arthur, living in separate parts of the same house in a rather tense atmosphere. Erwin arranged for the third floor of the Graz house to be made into a separate apartment, where the mother and child would come to live early in the new year of 1937. It would be a strange household. Anny spent most of the time in Vienna with her mother. Erwin confided to his journal some years later:

If you should ever want to write an epitaph or a memorial about me
and, as they say, my fate as a man . . . then you can say: Firstly, that a
woman who never loved me, on account of external circumstances
was in a situation to bear me a child, which is not so remarkable
because it happens every day. Secondly, that another woman loved it
more than if it were her own, it seems to me for the reason that it was
mine, which is not remarkable because under such circumstances (I
mean especially with the almost uninterrupted understanding of the
child's mother) it would occur roughly once in a thousand years.

Within a few weeks of beginning his lectures at Graz,
Schrödinger received a new honor, selection as one of the
foundation members of the Pontifical Academy of Sciences. In
January, 1936, Pope Pius XI had announced the formation of
the new academy – its purpose was to be the service of Truth.
Besides being scientists of world renown, the members of the
Academy were to be "men of irreproachable civic and moral
conduct who had always assumed a respectful attitude to
religion, without allowing a humanistic evaluation of strictly
scientific results to lead them to conclusions opposed to the
faith." The Vatican Secretariat of State, headed by Cardinal
Eugenio Pacelli, undertook to provide precise evaluations of
the moral qualities of those nominated for membership, and
some illustrious scientists were eliminated on the basis of this
scrutiny. The list of seventy members was published in October
1936. The physicists were Bohr, Debye, Keesom, Millikan,
Planck, Rutherford, Schrödinger, and Zeeman.

The inauguration of the Academy was held in Rome on June
1, 1937. The Holy Father could not attend owing to his serious
illness, but Secretary of State Pacelli welcomed the academicians, and conferred upon each of them the gold chain of
membership. Schrödinger's letter of thanks concluded: "Never
will I forget this visit to the Eternal City, a visit that was the
first for me and for that reason all the more impressive, and
never will I forget my profound obligation to demonstrate as
much as I am able my enthusiastic veneration toward a divine
institution which for all of us represents the most powerful
support against the frightful perils that are menacing human
culture."

SYLVESTER ABEND

On New Year's Eve, 1936, Anny was in Obergurgl for a skiing holiday. This was her fortieth birthday and she received a friendly greeting from Erwin, which she answered in a long letter:

I shall now use the rest of this Sylvester-Abend to send my thoughts to the two men who have played the greatest roles in my life . . . Your birthday letter to me . . . made me happy as it is the first personal emotion spoken from you to me for a long time . . . I believe, as you do, that this is the first step towards a change for the better in the relation between you and me. I believe you have understood very rightly wherein . . . a principal failure lies . . . You remember perhaps that I said many times: you cannot make any happiness while out of it you make me so unhappy . . . Erwin, believe me, I don't see fog and clouds at all, happily and confidently I see a future anyway. Even if the love between Peter and me should sometime come to an end, I would always be blessed that it had formerly existed, as I know that fate has given me the greatest happiness that a person can ever be given. And believe me, Erwin, I would not want to exchange this form of happiness for a marriage that allowed me always to be together with Peter. You yourself were the one who instilled in me that bourgeois marriage is definitely the finish of the deepest feeling of love because every day living destroys the magic . . .

Word for word I wish to accept the impression of your birthday letter. Gladly do I let my birthday today be a divide between the distressing past "and a happier life that insensibly passes through the shadows and is prepared for the bright daylight" . . . The best thing that life has taught me is unquestionably self reliance. That was for me very hard to learn, for my innermost and earliest wish was to be a part of another person whom I loved. Today I know that you were tormented by this above all, but I believe you did not know how difficult it was for me to free myself of it.

Turning to more practical matters, she reported that she had visited Innsbruck and found that Hilde was not making adequate preparations for the move to Graz, and she advised Erwin to light a fire under her if he wanted it to occur on schedule. Erwin was to come to Innsbruck about a week later to pick up Hilde for their own skiing holiday.

During these months Schrödinger was carrying on a bizarre feud with ICI about the disposal of the plumbing fixtures in the house he had rented in Oxford. At the beginning of his tenancy, he had paid £31 for certain fixtures on the understanding that when he left the sum would be refunded less an allowance for depreciation. ICI had taken a four-year lease on the house and after some delay they found a new tenant to whom they sold the fixtures for £20. They had a number of expenses in preparing the house for the new tenant, notably due to the neglected condition of the garden, and thus they applied the £20 to offset part of these costs. Schrödinger became furious at this "insolence" and sent complaining letters to Lindemann, Gordon, the British tax authorities, and of course to ICI. Lindemann hardly knew what to make of Schrödinger's behavior, in view of the generosity with which ICI had provided for him at Oxford. Eventually Alfred Mond, the chairman of ICI, sent him a cheque for £20 to conclude the matter, but it left a bad impression.

SCHRÖDINGER AND EDDINGTON

For the next few years Schrödinger's research was inspired by the cosmological theories of Arthur Eddington. Eddington (1882–1944) was the most distinguished astrophysicist of his time. His brilliant early work on stellar dynamics and on the structure and evolution of stars virtually created the discipline of theoretical astrophysics. The Astronomer Royal, Frank Dyson, pointed out that the eclipse of the sun on May 19, 1919, would afford an almost uniquely favorable opportunity to test a prediction of Einstein's general relativity, the bending of light rays from stars by the gravitational field of the sun. He organized two expeditions, one to Brasil and one to West Africa. Eddington went to Africa with E. T. Cottingham and they obtained the first confirmation of the Einstein theory.

After the advent of quantum mechanics in 1926, Eddington began to consider the deep and difficult question of how to bring together relativity theory, so effective in the cosmic domain, and quantum theory, so effective in the atomic domain – how to reconcile the physics of macrocosm and microcosm. From his middle years, he also devoted much time to the philosophy of science, and this study influenced his approach to problems of theoretical physics. The later Eddington was a lonely and controversial genius who combined numerical legerdemain and idealist philosophy in a prose style of such eloquence that it charmed even those who could not follow his reasoning. He believed that scientific knowledge is derived not from an external world but from the abstract structure of human thought.

While at Oxford, Schrödinger visited Eddington at Cambridge, where he was Plumian Professor of Astronomy and Experimental Philosophy. In 1936 he reviewed for *Nature* Eddington's *Relativity Theory of Protons and Electrons*. He was deeply impressed by the book but duly cautious: "We have here before us a sketch of unusual grandeur, of which not the details alone need further development and, maybe much modification. I am convinced that for a long time to come, the most important research in physical theory will follow closely the lines of thought inaugurated by Sir Arthur Eddingon." The simple but profound idea of Eddington was that the atomic structure of matter – the very existence of particles such as electrons and protons – is a consequence of the curved finite space of the universe devised by Einstein in his theory of general relativity. It is the finiteness of space that leads to discrete quantized energy levels and hence to atomicity. Schrödinger was enormously impressed by these ideas of Eddington. They extended his wave mechanics to the ends of the universe and gave his wave equation a cosmic significance.

A meeting of the Società Italiana di Fisica was to be held in Bologna, October 18 to 21, 1937, to commemorate the two-hundredth anniversary of the birth of Luigi Galvani, the discoverer of the electric current. Most of the great European physicists would be there: Bohr, Debye, Fermi, Broglie, Heisenberg, Hevesy, Kramers, Perrin, Oliphant, Richardson, Siegbahn, Aston, and Sommerfeld. They were invited to give papers based on their recent research. Schrödinger was invited to present a paper, but he decided to talk not about his own work but about "Eddington's Theory of the World."

The Bologna conference was officially opened by the puppet king of Italy in the Great Hall of the University. On October 20, news of the death of Ernest Rutherford was received, and Niels Bohr delivered a spontaneous eulogy of his old teacher – it marked the end of a great epoch in physics, during which the structure of the atom had been worked out in detail by theory and experiment.

Schrödinger began his paper with a theatrical metaphor: since the advent of general relativity, one can no longer regard space–time as the stage and matter as the actor; the properties of the one are so intimately related to those of the other that one can almost call them two different designations of the same thing. Thus when one applies quantum mechanics to an apparently isolated system, the mere fact of setting up the problem in space and time means that one must consider how the system reacts with the rest of the world. "One could say that the two theories in question are concerned with two complementary aspects of the world, one its connection on a large scale, the other its connection on a small scale . . . What we are missing is evidently the union of the two." Bohr must have felt a glow of satisfaction as he heard this tribute to his principle of complementarity.

Schrödinger presented the basic ideas of the Eddington theory in clear and unequivocal terms. He did not try to criticize or amend the theory. The members of the conference reacted with skepticism. On October 23, he wrote to Eddington:

I have just returned from Bologna, where I endeavoured to give a brief report of those parts of your theory which seem the most important to me. I met with an unvanquishable incredulity of the important group, Bohr, Heisenberg, Pauli, and their followers. I was in an extremely difficult position – spiritually I mean – because so many of your arguments are as ununderstandable to me as they are to them.

Despite his misgivings about Eddington's arguments, Schrödinger remained convinced that a quantum mechanical theory of the universe might be achieved. Thus, for several years, he devoted a major effort to problems derived from Eddington's cosmology. At the end of December 1937, he sent to the Pontifical Academy of Sciences a long memoir on "Proper Vibrations of Spherical Space." In the introduction to this paper, he noted that Eddington had considered the proper vibrations of a homogeneous, isotropic space of constant curvature, but had not provided any derivation of these wave functions. His paper provided mathematical results, which later he hoped to apply to the "treasure of ideas" contained in Eddington's work.

VIENNA VISITS

In addition to teaching at Graz, Schrödinger traveled to Vienna once a week for lectures and seminars, and he kept an apartment there. In 1932 Hermann Mark had moved to Vienna as professor of physical chemistry, and he often joined Erwin and Hans Thirring in various outings. He was about eight years younger than the other two, but since he was a professor that made no difference. The three men often discussed politics and Mark thought that both Hans and Erwin were socialists:

Thirring was a "red" as we would say at that time and Schrödinger disliked every politician and every government as a matter of principle. So they both disliked Dollfuss and hated Hitler . . . Neither Hans nor Erwin hated Schuschnigg, he was too colorless, he was a moderate, a philosopher, not a politician. He was no match for Hitler . . . The Nazis were very strong in Graz, in fact Schrödinger hated them there as much as he hated them in Berlin, and since they were much weaker in Vienna, he liked to come there to get away from them.

During the summer, Hermann, Hans, Erwin, Adolf Smekal, and a few others often went swimming in the Danube, changing into swimming trunks behind riverside trees. Sometimes the Thirring boys, Harald and Walter, would accompany them. Erwin was a strong swimmer and thought nothing of swimming across the river. They also would go in Mark's car to a big lake nearby, the Neusiedlersee. As Mark put it, "Those were happy days – but most people did not know they were dancing on a volcano."

Mark had good contacts with the Swiss chemists Karrer and Ruzsicka and they warned him in mid-1937 that Hitler was preparing to take over Austria. The Swiss papers were reporting that Mussolini had told Hitler that he would take no action if such a move was made. Mark told Hans and Erwin, "I shall leave Austria," they said, "For God's sake, why do you want to leave Austria, you are a professor." Mark responded "It's all going to go wrong." They said, "Impossible."

Erwin also had another interest in Vienna, for Hansi Bauer had returned from England to the family mansion. Her father had suffered a heart attack in March 1937, and he died later that year. Thus Erwin now had the problem of correlating his schedules with those of three women. Hansi and Erwin used to meet in the Vienna woods, where tiny villages and rustic inns provided ideal refuges for those who wished to escape the tumult of the city. Erwin loved the *Heuriger* (new wines) for which the local taverns were famous and loved even more the delightful company of the young woman who understood him so well.

ANSCHLUSS

Early in 1938, portents of disaster appeared in Austria. An intense Aurora lit up the sky so that in many districts fire alarms were sounded. The plague birds appeared on the streets of Vienna, albino sparrows with russet splotches like dried blood on their wings, and little girls were recruited by the Nazis to distribute pornographic broadsheets to passers-by.

On February 12, Hitler summoned Schuschnigg to his mountain hideout near Berchtesgaden and bullied him into signing over control of the police and foreign affairs to the Nazis in return for a promise of a guarantee not to invade. When Ciano, Mussolini's son-in-law, heard about it the next day, he wrote in his diary, "The Austrian chicken has fallen, or almost fallen, into the German soup pot."

At the height of the crisis, Schrödinger overcame his contempt for politics and made a cautious but unmistakable political statement to a huge audience. He was scheduled on Friday afternoon, February 18, to give a talk on "World Structure in the Large and in the Small" in the Auditorium Maximum of the University of Vienna. The hall was filled to capacity. Hans Thirring introduced the speaker as a "pathfinder of modern physics who has decisively accomplished the liberation of natural science from purely mechanical concepts . . . Although he has become world famous, he is a child of Vienna and has remained a true Austrian." In his lecture, Schrödinger first outlined the structure of matter at the atomic level, and then showed that the large-scale structure of the world is related to this small-scale structure, so that the ancient dream of the mystical *Naturphilosophie* of the middle ages, the unity of macrocosm and microcosm, was now becoming a reality. The nuclei of all the elements in the universe are composed of tiny particles, protons and neutrons, but the universe is not infinite in extent. If one imagined a pearl necklace made of neutrons and protons, which extended through all space and finally returned to its origin, the string would contain about 10^{40} particles. Since the whole universe contains about 10^{80} protons, Schrödinger triumphantly concluded that this shows how a property of the elementary particle determines the size of the universe. The spirit of Eddington was still strong within him.

After the scientific part of his lecture, he expressed two closing thoughts: "When one returns again from the kingdom of the stars to our world, one finds there a liking for a concept that wants to place one of the nations that live in this world over or under another one." At these words thunderous applause broke out, and only after some time could the speaker continue. "The science of inorganic nature compels a respect for a higher order that lies between pride and extreme modesty, an order which is over us and which always reveals itself to us in image and in likeness."

On March 6, Schuschnigg decided to hold a plebiscite on Austrian independence on March 13. Hitler was furious; he summoned his generals to prepare for an invasion on March 12. Schuschnigg was again cowed and called off his plebiscite. He asked the British government what to do. Ribbentrop, the German ambassador in London, invited Halifax, the foreign secretary, to tea, and explained that Hitler was only following a policy similar to that of Britain in regard to Ireland. Halifax advised Schuschnigg that he could expect no help. On March 11, Schuschnigg resigned and broadcast a plea to the citizens and armed forces to offer no resistance. In fact there was not a single gesture of resistance anywhere in Austria as the local Nazi bands took over.

Hitler arrived as a "tourist" to visit the grave of his mother in Linz. He was greeted everywhere with delirious enthusiasm. Actually, up to this point, he had not decided to take over Austria completely, but Goering now advised *Anschluss*. The takeover was announced on March 13 in the *Wiener Gazette* and the next day Hitler arrived in Vienna, which welcomed him with unfeigned rejoicing.

The sadistic behavior of the victorious Austrian Nazis in Vienna was worse than anything seen in Germany up till that time. Jewish shops and businesses were looted and anyone who resisted was beaten to the ground. Only the German military police kept the terror within any limits. For days a favorite pastime was to capture an elderly Jew and set him to work cleaning the sidewalks with an acid solution that severely burned the hands, while passers-by gathered to enjoy the spectacle. Within a few days there were 76,000 arrests in Vienna alone, and 6,000 dismissals from government offices and teaching posts. Thirring was one of the first to be sacked; storm troopers appeared and seized his office but he was not arrested. Mark was arrested, but fortunately one of his former students was in charge of the jail and Mark was able to obtain his freedom by payment of a large bribe to a top Nazi. He converted the rest of his funds into platinum wire, fashioned it into coathangers, and escaped to Switzerland. When the borders were closed, some who could not escape chose suicide. The Nazi control was enforced with many executions; no attempt was made to keep these secret and they were announced on large red posters throughout the cities.

Cardinal Innitzer sent greetings to Hitler and ordered all the Catholic churches to hoist swastikas and ring their bells in jubilation. He visited Hitler and promised him that the Catholics "would become the truest sons of the great Reich into whose arms they have been brought back on this momentous day." Hitler was delighted with the cardinal. The Lutheran bishops ordered services of thanksgiving for the *Anschluss* to be held in all the Protestant churches, so that the Christian prelates were in agreement that Hitler's coming was a blessing from God.

In Graz, there were fewer Jews to torture and Nazi control was already so complete that the *Anschluss* was less brutal than in Vienna. The University was forced to close and the Rector was dismissed. David Herzog, professor of Semitic languages and Rabbi of Graz, was seized outside his synagogue and thrown into the River Mur. Victor Hess was a native of Styria and a graduate of the University of Graz. In 1936, he had shared the Nobel prize in physics with Carl Anderson of Caltech. He was dismissed because he was a strong supporter of Schuschnigg, but he was able to leave the country for a professorship at Fordham University in New York. About fifty Jews in the medical faculty were summarily dismissed and many were imprisoned. Otto Loewi had received the Nobel prize in medicine in 1936 (with Henry Dale) for his discovery of the chemical transmission of nerve impulses across synapses. He was sixty-five years old and had been a professor at Graz for twenty-nine years, choosing to stay there despite many offers from more prestigious universities. He and his wife and two sons were thrown into prison. After two months he was allowed to leave for England after he had paid a ransom by transferring his Nobel prize money to a Nazi bank.

Schrödinger had planned to visit Oxford in the fall term of 1938, but early in March, George Gordon received a letter saying that "the great and important events we are living through in these days might make my presence necessary so that I could not get away." William Bragg told Gordon that he had heard that Schrödinger was in a concentration camp. Gordon wrote to Lindemann: "I propose to apply to Halifax to do something, since E. S. is still a fellow of the College."

Lindemann himself telephoned Halifax on April 6, and the foreign secretary promised to make inquiries through the British ambassador in Berlin, Neville Henderson, who was a great admirer of the Nazi regime.

APPEASEMENT IN GRAZ

Although Schrödinger was kept under surveillance, he was not molested in any way at this time, but he knew that if he wished to remain in Austria, he would have to try to appease its new rulers. The Nazi rector of the University of Graz, Hans Reichelt, had been assigned the task of compiling a list of the staff who were to be "cleansed." He advised Erwin to write a letter to the Senate setting forth his changed attitude. He did this and the letter was published on March 30. The Graz *Tagespost* gave it a prominent headline: "Confession to the Führer":

In the midst of the exultant joy which is pervading our country, there also stand today those who indeed partake fully of this joy, but not without deep shame, because until the end they had not understood the right course. Thankfully we hear the true German word of peace: the hand to everyone willing, you wish to clasp gladly the generously outstretched hand while you pledge that you will be very happy, if in true co-operation and in accord with the will of the Führer you may be allowed to support the decision of his now united people with all your strength.

It really goes without saying, that for an old Austrian who loves his homeland, no other standpoint can come into question; that – to express it quite crudely – every "no" in the ballot box is equivalent to a national [*völkisch*] suicide.

There ought no longer – we ask all to agree – to be as before in this land victors and vanquished, but a united people [*Volk*], that puts forth its entire undivided strength for the common goal of all Germans.

Well meaning friends, who overestimate the importance of my person, consider it right that the repentant confession that I made to them should be made public: I also belong to those who grasp the outstretched hand of peace, because, at my writing desk, I had misjudged up to the last the true will and the true destiny of my country. I make this confession willingly and joyfully. I believe it is spoken from the hearts of many, and I hope thereby to serve my homeland.

E. Schrödinger

The paper commented that there were many scholars in ivory towers who had been misled by the Dollfuss–Schuschnigg system, but the scales had fallen from their eyes as a result of the storm of enthusiasm that had swept the country. "The voice of blood calls also these men to their people [*Volk*] and thereby to find their way back to Adolf Hitler."

In later years, Schrödinger often regretted this groveling statement but the closest he ever came to an explanation was in a letter to Einstein (July 19, 1939):

I naturally knew there was a certain danger when I went back to Austria. But that the fortress would be surrendered without striking a blow, that I never reckoned until the end. Just a few days earlier I was with a section chief in the Ministry and said to him: if you put a rifle in my hand I will be glad to defend myself, but don't let me remain as a hostage in nazified Graz. You can imagine with what feelings just a few weeks after the overthrow I read the signature of the same gentleman under the orders of the new Minister! I hope you have not seriously taken amiss my certainly quite cowardly statement afterwards. I wanted to remain free – and could not do so without great duplicity.

On April 3, Karl Renner, the Socialist leader asked his followers to approve the *Anschluss*. On April 10, the Austrians voted in a plebiscite 99.73 percent in favor of the *Anschluss*, 4,453,000 yes, 11,929 no. The votes were cast under the eyes of Nazi storm troopers, so that probably no more than 90 percent of the Austrians really approved what they later called "Hitler's first conquest" or "the rape of Austria."

A brief notice of Schrödinger's attempt to appease the Nazis was published in *Nature* (May 21). His friends assumed that the letter had been written under duress and expected the worst. Much to their surprise, however, it turned out that Anny and Erwin were enjoying a peaceful spring holiday, skiing in the Tirol. W. J. M. Mackenzie, a member of Magdalen College, reported to President Gordon that he had met the Schrödingers there, and Gordon asked him to send Lindemann an account of this meeting, which he did in a letter of April 21:

We had a very long talk about things in general . . . The following points may be of interest: (1) Schrödinger himself is in good health, has not at any time been under arrest, and has not been personally

molested; Mrs. Schrödinger is still convalescent after an operation
but seems to be getting on well. Their correspondence however is
probably being tampered with . . . His personal position is one of
complete uncertainty, for better or for worse; he summed it up by
saying that they ought logically either to promote him to a better job
or put him in a concentration camp, but he had no idea which it
would be . . . In regard to future plans; he has clearly no feeling that
it is necessary for him to get out of the country at all costs as soon as
possible, and would on the whole like to make his peace with the
regime if they will let him.

Schrödinger's remarks about "promotion to a better job"
could have referred only to the Vienna professorship from
which Thirring had been dismissed. In his political naivety he
did not realize that all such vacancies would be filled by eager
and dedicated Nazis. He did begin to worry about his Swedish
securities and undertook a delaying action to avoid their transfer to a German bank.

On April 23, an international meeting of the German
Physical Society was held in Berlin to celebrate the eightieth
birthday of Max Planck. Max and Marga were delighted to see
Erwin and Anny at the social events that accompanied the
scientific sessions. By this time the German physicists had
adapted themselves well to the Nazi regime, and the absence of
Jewish colleagues would have been hardly noticed. They were
probably in general pleased that Schrödinger, the one former
exception to their compliance, had now rejoined the fold.

Schrödinger had ventured into the center of Nazi power and
returned uneventfully to Graz, but he found an unpleasant
surprise awaiting for him there. On April 23, he had been
summarily dismissed from his honorary professorship at the
University of Vienna, in a communication from the office of
the Dean of the Faculty of Philosophy.

His appointment at the University of Graz, however, was
not affected at this time, since the Rector had taken personal
charge of the purging of the university staff. The Senate voted
to change the name of the university to Adolf Hitler University, but the Senate was itself abolished before this change
became effective. The University of Graz became a major

center for Nazi education, with special courses in racial studies, war chemistry, and for training the SS medical corps who would operate the extermination facilities. After the defeat of Hitler, the Nazi staff fled, but within a few years most had returned, and the university archives were sealed so as to prevent any revelations concerning the Nazi period.

Bad news travels fast, but in the case of Schrödinger's dismissal from Vienna, it surpassed itself. On the same day that the letter of dismissal was sent, Eamon de Valera wrote from the Office of the Taoiseach (Prime Minister) in Dublin to E. T. Whittaker in Edinburgh:

In an evening paper a couple of days ago I saw it noted that Professor Schrödinger had been dismissed from his post. I suppose that it has not been possible for you to get in touch with him? I am very anxious that we should secure his services in connection with the project we discussed when you were here. [Establishment of an Institute for Advanced Studies.] If you are able to communicate with him, will you please convey to him an invitation from me to come to Dublin. Whilst we are waiting to have the scheme worked out, some special financial arrangement can be made for him.

LAST DAYS IN GRAZ

The situation of the Schrödingers in Graz was more precarious than they realized. On May 9, Lindemann received a letter from Halifax:

My dear Lindemann,
    With reference to your telephone message of the 6th April I am writing to let you know that our ambassador in Berlin has now had a reply from the German Minister for Foreign Affairs to the enquiries he made at my request about Professor Schroedinger.
    I am sorry to have to tell you that Herr von Ribbentrop's reply is unfavourable. According to the German authorities Professor Schroedinger left Germany in 1933 for political reasons and, after a brief stay in Oxford, settled in Graz where he proceeded to busy himself as a fanatical opponent of the new Germany and of National-Socialism. A recent examination of his case had shown that Professor Schroedinger had, up till very lately, remained in constant communication with German émigrés abroad and it was therefore feared that permission for him to visit Oxford would merely offer him a further

opportunity of resuming his anti-German activities. In these circumstances Herr von Ribbentrop fears that a further course of lectures by Professor Schroedinger at Oxford could only serve to harm Anglo-German relations and is therefore, to his regret, obliged to refuse my request.

I much regret that my intervention on behalf of Professor Schroedinger should not have been more successful but I fear that, in view of the attitude adopted by the German authorities, there is nothing more that I can do.

> Yrs. sincerely,
> Halifax

Richard Bär was an old friend of the Schrödingers from Zurich days. He had been using his family wealth to relieve the distress of refugees from Nazi Germany. On June 10, Bär and his wife met Anny in Constanz. He reported to Franz Simon in Oxford: "Erwin is no Nazi, despite the ominous letter that he freely wrote . . . He would now like to live in peace with the regime, like Laue and Heisenberg. He hopes to get leave for the fall semester to come to Oxford." Anny had an important reason for being in Constanz, but Bär did not mention it in his letter to Simon.

Simon replied to Bär with some political realism.

If Schrödinger comes out, which appears to me doubtful, in my opinion he can never go back again. The situation is thus: the letter made an awful impression here. In order to restore his reputation – and in a certain sense that of all the emigrants – it has been spread abroad that he wrote this letter only under extreme pressure. When he now arrives here, he must either say that he has written it voluntarily, then he is in a really unpleasant situation, especially since people know very well what he thought before and what he just a little while previously expressed in writing to local colleagues. Or, he says that he was forced to do it, then he can no longer go back. (There are always informers here who make reports to Germany.) I wanted to make this clear to you, so that in case you have an opportunity you can pass it on to Schrödinger. I might mention that in this connection the famous saying of the King of Hanover is becoming cited here: Professors can be bought like whores.

Anny later told the story of the reason for her trip to Constanz as follows:

Well, when the Nazis came to Austria . . . it was really absolutely like a prison . . . De Valera knew that we were in danger and he let us know that there was a possibility for an institute for advanced studies in Dublin which he wanted to create if my husband said, in principle, "yes I will come." It was absolutely sure he could not write to my husband because everything was censored. So he asked [Whittaker] to ask Born; Born wrote to our friends in Zurich [Bär]. Our friends in Zurich told a Dutchman, who came to Vienna, about the possibilities. We were not in Vienna. He didn't come to Graz, he came to my mother and told her this important thing. She was afraid to take such a very important message without having anything written down, so she wrote just down in a few lines that de Valera wanted to create an institute for advanced study and whether he would come, in principle. This little piece of paper my mother sent to Graz. We saw it and we read it three times and then destroyed it, threw it into the fire, and told nobody about it at all. I went with my car as far as Munich. Thirring went with me and I went to . . . Constanz; I met our friends there and I told them, "Yes, in principle, he will come. But nothing should be done that will let anybody know that we are going away" . . . My friends wrote to Born, and Born told Whittaker, and Whittaker told de Valera, and that was finished. He never spoke to de Valera [before all this]; he never knew de Valera – nothing at all.

FLIGHT FROM GRAZ

Schrödinger was oblivious of the gathering storm of Nazi vengeance that was about to overtake him. Shortly after his return to Graz, he received the following curt notice from the Ministry of Education:

Vienna, 26 August, 1938
To: Herr Professor Dr. Erwin Schrödinger
On the basis of Sect. 4, Paragraph 1 of the ordinance for renovation of the Austrian civil service of 31 May 1938, RGBI.I S.607, you are dismissed. The dismissal is effective as of the day of arrival of this notice. You have no right to any legal recourse against this dismissal.

The reason for dismissal specified in the ordinance was "political unreliability." After this action by the Austrian authorities, the Berlin University stripped Schrödinger of his title as Professor Emeritus and ordered that his name be stricken from all the records of the university.

In dismissing Schrödinger, the Nazis did not act in rational self-interest, since it would have been a propaganda success to exhibit as a convert to their cause the only world-famous German scientist (without a Jewish family) who had spurned them.

Schrödinger was dismayed by his unexpected dismissal and he went immediately to Vienna to talk with an official high in the government. He was told not to worry because "It is easy for you with your name to get another job in industry, or somewhere." His friend Hans Thirring, who had been dismissed earlier, managed to survive with the help of consulting jobs, but he had some friends in high places. Schrödinger said in effect, "Well for a theoretical physicist that is not so easy, and look what happened to all the Jews, their papers became invalid or were misplaced so they could not get work. I shall have to go to a foreign country to find my living again." The official said "They won't let you go to a foreign country. Have you still got your passport?" This was apparently well meant; it was not a threat, but it gave Schrödinger a shock because he never thought he might be prevented from leaving Germany.

Now he knew that there was no time to waste. He rushed back to Graz and in three days they were ready to leave. Everything they could take was packed into three suitcases. It was forbidden to take money or valuables out of the country, so they left almost all their possessions in Graz, including the gold Nobel medals and the chain of the Papal Academy. They bought two return tickets to Rome. They left Graz on September 14 with ten marks in their pockets.

As Anny continues the story, "We didn't have the money to pay the porter in Rome, so we had the taximan pay the porter, and the hotel pay the taxi." Fermi met them and lent them some money. He warned them: "Don't write from Rome because it is already dangerous – it might be censored." From the Papal Academy, which is beautifully situated in the Vatican Gardens, Schrödinger wrote three letters: one to Lindemann to tell him that they had left Graz, one to Bär to ask him to send some money, and the third to de Valera who was the President of the League of Nations in Geneva. The letters were posted from Vatican City on Saturday. On Monday Schrödinger talked with de Valera for the first time. He said he was very glad that they were out of Austria and they should come to Geneva to discuss a few things, but as soon as possible should go on to England or Ireland because there was such a great danger of war. It was near the time of the Munich conference and the betrayal of Czechoslovakia. It was forbidden to take any money out of Italy, and the Irish consul gave them a pound each and first-class tickets to Geneva.

"I was quite happy," Anny recalled:

I felt already safe but my husband didn't feel safe at all. At Domodossola, they looked at our passports, but the luggage hardly at all. The passports were all right. But before we came to Iselle – at one end of the Zipplertal [Simplon tunnel] – a carabiniere came into our compartment and we had to leave with all our luggage. He had a piece of paper with our name written on it. It was really the fright of my life. We had to take off everything [from the train]. [Erwin] was separated from me with the luggage and I was in another place. I couldn't speak Italian; there was a woman who looked through my things, saying, "Put everything toward the X-rays." They X-rayed every single bit of my things – my handbag, my teeth, everything. After about half an hour, or three-quarters of an hour – the train had to wait; it was an express train – we were allowed to enter the train again . . . They had looked at our passports and we had visas for all of Europe because de Valera had said that we can't go through France, we must go through Spain or Portugal or something. We had everything. We were asked if we had some money. We said that we had one pound. Then they thought that we had to be smuggling something, because one can't go through Europe on one pound.

De Valera was very, very pleased when we arrived. He was already in full dress because they had a banquet in the evening, but he received us and he was very kind, and we stayed for three days in his hotel and then went on.

RETURN TO OXFORD

During the month of September 1938, war in Europe seemed inevitable as Hitler moved quickly to take over the northern borderlands of Czechoslovakia. Chamberlain and Daladier presented the Munich decisions to the Czechs and told them they must surrender. Chamberlain returned to 10 Downing Street to the cheers of his many supporters, to whom he brought "peace in our time." Churchill was one of the few skeptics: "We have suffered a total unmitigated defeat."

When Lindemann received Schrödinger's letter from Rome, he became quite angry: "Schrödinger asks whether he can come again to Oxford. Is he mad? Doesn't he realize after this letter he has published what people think of him?"

Then suddenly, in early October they appeared. They came to the Simons and it was an uncomfortable evening for everyone. Franz Simon told Erwin, "Well, we have tried very hard to tell people here that this letter was written under duress. We don't know the conditions, we said, probably somebody with a gun stood behind him and said 'you sign this.'" First he said "What letter?" and then he became quite excited and said "What I have written, I have written. Nobody forced me to do anything. This is supposed to be free country and what I do is nobody's concern." The Simons told him with more emotion than logic that he had made the situation more difficult for all the other refugees. They said that if war broke out tomorrow, he would be in serious difficulties. Erwin said he was grateful that they had made him aware of the matter, but it concerned only himself. Anny supported Erwin strongly – whatever he did was right. The next day he saw Lindemann who told him the same thing as Simon.

Schrödinger wrote a long letter to Born, which concluded. "Ach, if only one could make it clear to these English how it really seemed over there! With them [the Nazis] you take care. You do not believe anyone." Born commented to Simon, "How are you supposed to believe a man who has published that pretty letter?"

On October 7, Simon wrote to Born, with more details about Schrödinger: "He has in this matter a frightfully bad conscience, but does not have it in his heart to confess that what he has done is wrong (which does not surprise me in someone who so loves himself to the exclusion of all others) . . . One must unfortunately say that he behaves like a spoiled boy, besides this also with his immature sexual complex . . . These happenings have disturbed us greatly."

Max Born does not appear to have replied to this letter. His friendship with Schrödinger survived the episode of the unfortunate letter. Max understood that the sources of political depravity lie deep in the human condition, so that no person can claim immunity. He was better able than Simon or Lindemann to appreciate the volatile character of Schrödinger, who did not feel deeply about any political questions, and who honestly could see nothing wrong in writing an expedient letter to a bunch of contemptible gauleiters. Erwin had never claimed or aspired to be the conscience of the German physicists, and the idea of going to a concentration camp as a political protest had no attraction for him. In retrospect, harsh criticism of Schrödinger was hardly appropriate from those who were safely ensconced far away from the Gestapo.

The Schrödingers stayed two months in Oxford as "paying guests" with John and Barbara Whitehead at their house, 22 Charlebury Road. John was a brilliant mathematician and fellow of Balliol College. He had taken his doctoral degree at Princeton with Oswald Veblen. His father was the Anglican Bishop of Madras and a brother of the philosopher Alfred North Whitehead.

In August, Hansi Bohm, by a mixture of bribery and good luck, had managed to escape from Austria to England via Switzerland. Sometimes Erwin visited her small furnished flat in Hampstead. They wandered together through the neighborhood streets, stopping in the cheap teashops of pre-war London, where awful sugar buns were displayed in fly-specked windows and puddles of tea were seldom wiped from the plastic tables. The conversation was always spirited as the lovers, by now completely at home with each other, lapsed into their native Viennese. Erwin would laugh at Hansi in the kitchen, saying she looked like a Burgtheater actress trying to play the part of a cook.

BELGIAN INTERLUDE

De Valera was preparing the legislation necessary to establish the Institute for Advanced Studies, but it would require at least a year to work its way through the parliament. Meanwhile he arranged for a grant of £200 through University College and Schrödinger went over to Dublin on November 18 and stayed for several days, discussing future plans with de Valera and Arthur Conway, Professor of Mathematics.

When he returned to Oxford, he was delighted to find a letter from Jean Willems, Director of the Francqui Foundation, which offered a visiting professorship at the University of Gent for the academic year 1938–39, with an honorarium of 75,000 Belgian francs ($10,000). Schrödinger immediately accepted the appointment, and wrote: "as regards the details of the letter, I am, of course, enchanted since they meet my intense desire of being given again an opportunity to display some external efficiency and to do useful work." In early December Erwin and Anny visited Paul and Margit Dirac in Cambridge and then traveled to Belgium, arriving about December 15.

From the beginning of the second term, Schrödinger gave two lecture-courses. An evening seminar in the department of Professor Verschaffelt was devoted to topics in molecular statistics. Visitors came from Brussels and other universities and the discussions sometimes continued late into the night, with "indefatigable interest . . . in the advanced and subtle questions that were presented." He also lectured in Brussels, Louvain, and Liège. In Louvain he came to know the famous cosmologist, Georges Lemaitre, and they had extensive discussions of the theoretical analysis of possible models of the universe.

During this time Schrödinger wrote one of the most revealing documents that we have from his pen. It is an untitled sheaf of six handwritten pages attached to a few lines of Spanish verse suggested by Calderón, "Gustas y disgustas no son mas que imaginares."

> No es consuelo de desdichas
> Es otra desdicha aparte
> Querer a quien las padece
> Persuadir que no son tales.

It is no solace of misfortune / It is another misfortune besides / To ask the one who endures them / To be persuaded they are not such.

"Every second of our lives," he wrote, "is saturated with the physical consequences of science or, as we could say, with excrements from the progress of research." Schrödinger was not the first, nor the last, to suggest that technology has caused a deterioration in the quality of man's relation to the deeper sources of his being. Aldous Huxley, like Erwin a true believer in Vedanta, complained that: "Modern man no longer regards Nature as being in any sense divine, and feels perfectly free to behave towards her as an overweening conqueror and tyrant."

When Schrödinger performed scientific work, however, he had a full awareness of the divinity of Nature, so that his approach to research was in accord with his belief in Vedanta. He expressed it as follows:

Science is a game – but a game with reality, a game with sharpened knives . . . If a man cuts a picture carefully into 1,000 pieces, you solve the puzzle when you reassemble the pieces into a picture; in the success or failure, both your intelligences compete. In the presentation of a scientific problem, the other player is the good Lord. He has not only set the problem but also devised the rules of the game – but they are not completely known, half of them are left for you to discover or to determine. The experiment is the tempered blade which you wield with success against the spirits of darkness – or which defeats you shamefully. The uncertainty is how much of the rules God himself has permanently ordained, and how much appears to be caused by your mental inertia, while the solution generally becomes possible only through freedom from this limitation. This is perhaps the most exciting thing in the game. For here you strive against the imaginary boundary between yourself and the Godhead – a boundary that perhaps does not exist. To you may perhaps be given the freedom to unloose every bond, to make the will of Nature your own, not by breaking it or conquering it, but by willing it also.

The grave error in the technically directed cultural drive is that it sees its highest goal in the possibility of achieving an alteration of Nature. It hopes to set itself in the place of God, so that it might force upon the divine will some petty conventions of its dust-born mind, and it overlooks the possibility of reaching its goal in a way that expands the divine sparks to an indistinguishable image of the divine will or the divine action (which is one and the same).

The God–Nature wills everything that it does, scarcely everything that it wants, but otherwise nothing. This wonderful self-limitation is not really one, because outside of it nothing is possible which it has renounced, just as spherical space is finite without ever pushing against boundaries that separate it from anything else. This is what we worship in the Godhead. To become like it, we must realize it, not seek our own fame like naughty children. Only when we choose this way does everything maintain the character of the game, with that absence of gravity which is so characteristic of every higher spiritual activity.

Erwin thus found in the philosophy of Vedanta his understanding of the nature of scientific research. He may have been thinking of one of his best loved quotations from the Bhagavad Veda:

> Who sees the Lord dwelling alike in all beings
> Perishing not as they perish
> He sees indeed. For, when he sees the Lord
> Dwelling in everything, he harms not self by self.
> This is the highest way.

Lemaitre had shown, as had the Russian Alexander Friedman, that a static Einstein universe is not stable and must either expand or contract. At this time, Schrödinger published a paper, "The Proper Vibrations of the Expanding Universe," which contains an important discovery: the first indication that an expanding universe under certain conditions may require the creation of matter. About ten years later, Hermann Bondi and Fred Hoyle developed a steady-state cosmology based on continuous creation of new matter in an expanding universe; models of this kind, however, are no longer considered to be probable.

On May 1, the Rector of the University of Gent informed Erwin that the Academic Senate had recommended him for an honorary degree, the Doctor of Sciences. This was his first honorary degree. It was to be formally conferred at a ceremony on October 9, but the outbreak of war made this impossible.

In the late spring, they moved from Gent and settled in the seaside resort of La Panne, 7 Sentier des Lapins (Rabbit Lane). Arthur March brought Hilde and Ruth to Belgium and returned to Innsbruck after a brief visit. Thus when war broke out on September 1, Erwin's unusual family was reunited. In

Innsbruck, March became active in the Tirolean underground resistance to the Nazis.

On September 29, Schrödinger wrote to the Rector to say farewell and to thank him for

the kind hospitality, that I have enjoyed at the University of Gent, and to express to you how deeply indebted I feel to your fatherland for the refuge and help that it unreservedly granted us in critical times. All this remains written in my heart; I should be happy if I could ever repay even a small part of it . . . We travel next week for the time being to Ireland where I have for the autumn a temporary professorship. One is cured of further worries in a time that has made almost every individual fate in Europe dependent partly on the outcome of disturbances that shake it up completely, partly on any gust of wind that deflects a musketball by a handsbreadth.

There was a problem in obtaining a visa for England, since they were now technically enemy aliens. With the help of de Valera, the Schrödinger party was granted a twenty-four transit visa, and they all arrived safely in Dublin on October 6, 1939.

CHAPTER 10

Wartime Dublin

The establishment of the Dublin Institute for Advanced Studies was due almost entirely to the efforts of Eamon de Valera, the Taoiseach (prime minister) of Ireland. Dev, as he was called by friend and foe, was born in New York City in 1882, so that he was five years older than Schrödinger. His mother was an Irish servant girl and his father a Spanish artist. Eddie, their only child, was three years old when the father died, and he was brought to Ireland to be raised by his grandmother in a cottage in County Limerick. As a schoolboy, mathematics was his best subject and he never lost his love for it.

Dev was commandant of one of the Irish battalions in the Easter Rebellion of 1916. The British destroyed the center of Dublin with incendiary shells and forced the republican soldiers to surrender. The British commander ordered summary courts-martial and executions of all the leaders of the rebellion. Dev had been sentenced to death one day later than the commandants in the city center, and when the British government halted the executions, his life was saved and he was sent to Dartmoor prison.

In 1924, he founded a new party, Fianna Fail (Band of Destiny) to pursue republican principles by electoral means, and in 1932 he achieved power in coalition with a small Labor contingent. In 1937, he won an absolute majority, scrapped the oath of allegiance to the British crown, and secured the adoption of a new constitution for Eire, the Irish Republic, without, however, completely severing ties with the Commonwealth.

DUBLIN INSTITUTE FOR ADVANCED STUDIES

Even in the midst of his political work, Dev never forgot his intellectual loves, mathematics and the Irish language. He noted the foundation in 1930 of the Institute for Advanced Study in Princeton, and the fact that Einstein and other scientists had gone there as refugees from Nazi Germany. Abraham Flexner had described the ideal of such an institute:

It should be a haven where scholars and scientists may regard the world and its phenomena as their laboratory, without being carried off in the maelstrom of the immediate; it should be simple, quiet, comfortable, quiet without being monastic or remote . . . Its scholars should enjoy complete intellectual liberty, and be absolutely free from administrative responsibilities or concerns.

Dev began to consider whether a similar institute might be established in Dublin. He thought that it could provide a common ground for scholars from the University of Dublin (Trinity College Dublin) and the new National University of Ireland (University College Dublin). His plan was to begin with two Schools, Celtic Studies and Mathematical Physics. The "humanist quality and national flavour," of the former would be balanced by the scientific precision and international character of the latter. He consulted Edmund Whittaker, who stressed the importance of securing a physicist of world renown, and suggested that Schrödinger was under pressure since the Nazi seizure of Austria and might be persuaded to come to Dublin. De Valera, an expert in undercover activity, immediately arranged the secret contact with Schrödinger that led to their meeting in Geneva in September 1938.

The bill to establish the Institute was introduced in the Dail on July 6, 1939. Since Fianna Fail had an absolute majority, there was no doubt about its eventual passage; but the opposition attacked de Valera fiercely, saying he was trying to satisfy his vanity with a pretense of scholarship at a time when war in Europe was imminent and Ireland faced military and economic perils.

ARRIVAL IN DUBLIN

Dublinn is an Irish word meaning "a dark pool," but the official name of the city is Bhaile Atha Claith, "the town of the ford of the hurdles," after the site of a shallow crossing of the River Liffey. Another etymology was given by an Englishman, John Head, in 1600: "Many of its inhabitants call this city Divlin, quasi Divel's Inn, and very properly it is by them so termed; for there is hardly in the world a city that entertains such devil's imps as that doth." In the ensuing centuries, the city's reputation gradually improved, but when the Schrödingers arrived it was still known as "dear dirty Dublin," and the living conditions of its poor were among the worst in Europe. It was, however, again the capital of a free country, had a population of half a million, three cathedrals and six theaters (not counting the popular movie palaces). The central area of O'Connell Street had not yet recovered completely from the devastation of Easter Week, 1916, and memories of the civil war had left a bitter residue in Irish politics.

On October 7, 1939, Erwin, Anny, Hilde, and Ruth arrived in Dublin. One of the first friends they made was Albert J. McConnell, always called "A. J.," who, as professor of mathematics at Trinity, was involved in planning for the Institute. They came around to his apartment and discussed the problem of finding a permanent place to live. Erwin displayed his best Viennese charm, but he made it clear in no uncertain terms that Mrs. March was to be treated on exactly the same terms as Mrs. Schrödinger.

The family wanted to live near the sea, and they soon found a house in Clontarf, a fashionable suburb on the northern shore of the Bay of Dublin. The Schrödinger home was at 26 Kincora Road, a semi-detached two-storey brick house in a row of similar dwellings on a pleasant tree-lined street about 300 meters from the seashore, where the Wicklow mountains form a hazy blue background across the waters of the bay. They were to live here for seventeen years. It was not an elegant residence as compared, for example, with the one they had rented in Oxford, but it had adequate room for the family, a small garage at the side, and a pleasant walled garden at the back. Erwin was not too pleased with the construction and the typically British plumbing and heating. After renting for several years, he bought the house for £1,000 in 1943, selling it in 1956 for £2,150.

The house was about 6 kilometers from the Institute in Merrion Square. Erwin, an expert cyclist, had no difficulty in making the trip in all weathers. He often wore a waterproof cycling suit and his jaunty beret, and he soon became a familiar figure among the many cyclists of wartime Dublin.

Erwin's unusual family did not cause much adverse comment in Dublin. Ireland was officially a puritanical society, under the strong Jansenist tradition of the Irish Church. At that time, the *Irish Catholic* was upset by women wearing slacks because "their stylish cuts throw the feminine figure into undue prominence." Other ecclesiastics were troubled by schoolgirls who crossed their legs in public although "they had presumably been educated by the nuns in Christian modesty." There was a rigid censorship that tried to exclude not only most of contemporary European literature but even collections of Irish folktales that contained mild improprieties. At intervals, lists of forbidden books were printed in the *Irish Times*, and they could be ordered from England.

The official puritanism was alleviated by an informal spirit of *laissez-faire*. As Erwin once remarked, "In Germany, if a thing was not allowed, it was forbidden. In England if a thing was not forbidden, it was allowed. In Austria and Ireland, whether it was allowed or forbidden, they did it if they wanted to." Thus, in reality, Dublin was not a strait-laced city. For example, the inseparable actor-managers of the Gate Theater, Micheal MacLiammoir and Hilton Edwards, always called "the boys," gathered only glances of admiration as they strolled about together painted and powdered, and Brendan Behan, on the run from the Garda, could find a refuge in many a respectable house.

The situation caused by the war in Europe was always officially called "the Emergency." The government declared a policy of strict neutrality and placed its small army on a war footing, expecting an invasion by England or Germany, or both. In the first year or so of the war, there were few hardships. An endless series of letters in the *Irish Times* protested the persistence of daylight saving, and there was concern about the loss of stud fees for Irish stallions. Otherwise life went on very much as before. Gradually, however, as the U-boats disrupted shipping, conditions became more difficult. There was always plenty of food, although butter and sugar were eventually rationed. A visitor from England was astounded by the sumptuous helpings of roast beef at dinner at the Maynooth Seminary. The most grievous scarcity was of tea, sometimes cut to half an ounce a week per person, but you could always get a cup in a tearoom. There was never any shortage of alcoholic drink. The bread became a wholesome, soggy brown loaf, made from local wheat mixed with barley and oats.

As the war went on, there were serious shortages of coal and oil. Some horse-drawn buggies appeared, but most people used bicycles. Household gas supplies were strictly limited and special police, called "glimmer men," were empowered to enter any house to check the stoves for compliance. Remembering Vienna in 1919, Schrödinger found small reason to complain about wartime Dublin. There were no tourists. In the words of John Ryan, "the country was clean, uncluttered, and unhurried. All in all, it was not the worst 'Emergency' we ever had."

DUBLIN 1939–40

Soon after he arrived in Dublin, Erwin met Monsignor Patrick (Paddy) Browne who was then professor of mathematics at St. Patrick's College, Maynooth, where most Irish priests received their training. He had been actively involved in helping de Valera plan the Dublin Institute. Many people say that Paddy Browne became Erwin's best friend in Dublin, despite some serious differences they had from time to time. He was a Rabelaisian priest, in the literal sense that he knew much of Rabelais practically by heart, and also in the figurative sense that he had a great store of bawdy stories. He was also rather a gargantuan man, almost 2 meters tall, with a leonine head and bushy hair, so that he stood out in any group, and towered over Erwin. Browne was an excellent teacher of mathematics but not a researcher in the subject; he was also a classical scholar and a Gaelic scholar. His elder brother Michael became the head of the Dominican order and was made a cardinal, but Paddy was too broad a man to win preferment in the conservative Irish church. Erwin's classical education was no match for Paddy's, and Paddy's mathematics was not in the same class with Erwin's but they respected and loved the intellectual qualities they found in each other. They were also both poets by avocation.

Schrödinger's first public appearance in Dublin was on Wednesday afternoon, November 14, when he read a paper before the Dublin University Metaphysical Society on "Some Thoughts on Causality." The talk was well publicized and attracted a large audience, by no means restricted to philosophers. He reviewed several themes that he had considered in earlier works: (1) the impossibility of prediction even in classical physics due to experimental uncertainties in the initial conditions, (2) the teaching of general relativity that matter is an inherent property of space and time and hence is not something predictable in space and time, (3) the uncertainty principle of quantum mechanics, which introduces indeterminism into the behavior of microscopic objects. He concluded that causality is not a necessity for logical thought, but he advanced this view tentatively, with a quotation from Miguel Unamuno to the effect that "a man who succeeded in never contradicting himself was to be strongly suspected of never saying anything at all."

This scientific dissection of causality did not please every philosopher in Dublin, since it contradicted the official Catholic philosophy based on Aristotle and Aquinas, which purports to trace an uninterrupted chain of cause and effect back to the first cause, which is God. Thus, strictly speaking, Schrödinger may have been guilty of heresy; since he spoke at Trinity College, this would not have been considered remarkable, but an interesting repercussion would occur a couple of years later.

In November, Schrödinger began an informal course of lectures on wave mechanics for undergraduates at UCD. His excellence in lecturing soon became famous. He used a black-board with rare artistry – not for him the scribbled equation hastily erased or the words addressed to the board and imper-fectly reflected at the audience. The written material appeared in logical order as if in accord with a mental map of the blackboard surface, and his words were spoken in his clear, light, precise English. He wrote a notice about these lectures "for the press (if any): A course of lectures on the latest form of quantum theory intended for advanced students in physics was started yesterday in UCD at 3 p.m. by the Viennese professor E. Schrödinger, Nobel prize winner, honorary member of the Royal Irish Academy, honoured by the Nazi government with pensionless dismissal, without notice, from his academic chair in Austria." The first lecture will consider "two discomforting features in the Planck–Bohr theory: (1) the quaint basic assumption about the discontinuity of states, (2) the frequency of spectral lines entirely different from the frequency of atomic oscillations." Thus Schrödinger brooked no delay in giving Dubliners his views on "those damned quantum jumps."

On December 11, he presented his first Dublin research paper to a meeting of the Royal Irish Academy, and it was published in the *Proceedings*. It was the first of several papers on a new and convenient way of solving the Schrödinger equa-tion. He had been an honorary member of the Academy since 1931. Most of his papers were henceforth to be published in its *Proceedings*, where they were elegantly printed and appeared without any delays. The volume of research from the Institute eventually became so great that the Academy was financially hard pressed to publish it all, and the Institute was asked to pay part of the costs.

IRISH HOLIDAYS

In the first week of May 1940, as springtime came to Europe, there was little activity on the battlefields of what was being called the "phoney war." The long Whit Sunday weekend was coming, and Erwin took off for a short cycling holiday: "It was my first spring in Ireland. I had got myself and my bike on to the train to Galway. The high sombre walls, which still today girdle the seats of the old country lords, recede and disappear. Get-at-able nature clambers up close to you; calm, silent, unstinted meadows, poplars, willows, grazing beasts. Not exciting romance, but so very soothing." But on Sunday evening his feeling of peace was destroyed and he rushed back to Dublin when he learned that Germany had invaded the Low Countries. By June 3, the Nazis had overrun Belgium, the Netherlands, and defeated the French army. The British forces were being evacuated through Dunquerque.

Paddy Browne had a house at Dunquin, on the extreme south-western tip of the Dingle Peninsula in County Kerry, just a few hundred meters from steep cliffs overlooking the Atlantic Ocean. It was a wild, Gaelic-speaking country of sea and mountains. Erwin and Anny were invited to visit there towards the end of their first summer in Ireland. The three children of Paddy's sister Margaret were there at the time, Maire (eighteen), later Mrs. Conor Cruise O'Brien, Seamus (sixteen), and Barbara (twelve). The monsignor had practically raised the children since their father Sean MacEntee, who was now in Dev's government, had been so occupied with Irish revolutionary politics. Paddy Browne had been a student at Heidelberg and he taught the children a few German folk songs. They became very fond of Anny who knew a number of unfamiliar ones – "Phyllis ging wohl in den Garten," "Horch was kommt von draussen 'rein." Erwin was less popular since he wrote a poem for Barbara about neatness and good behavior. It began "You're growing a lady now, my dear, For instance, clean your nails." Such an equation of personal grooming with virtue was incomprehensible to the MacEntee children, and it created "an absolute cultural barrier."

Despite her unkempt nails, Barbara was a beautiful child and Erwin became infatuated with her. She was the third instance of his "Lolita complex," taking her place along with "Weibi" Rella and "Ithi" Junger. Perhaps it was the "little boy" aspect of his nature that attracted him to adolescent girls, but to him these loves seemed like erotic arrows that pierced his soul unbidden and unexpected. The situation became so incongruous that someone, probably Paddy Browne, had a serious word with him, and muttering imprecations, Erwin desisted from further attentions to Barbara, although he listed her among the unrequited loves of his life.

On another visit to Kerry, the following year, Erwin and Anny went on a bicycle excursion from Dunquin, and had an adventure which was later related by Sean O'Kennedy who was on a hike with two teenage pals:

It happened that three of us (all natives of the district) were bantering by the road-side when we observed the pair, who were then altogether unknown to us, cycling in our direction. Before passing, we advised them to dismount the bikes as they were almost certain to have punctured tyres if they cycled over an extensive patch of jagged stones immediately ahead. They paid no attention, perhaps because we appeared to be loungers who were up to no good and to them the patch of stones posed a lesser danger. In the event, she got through without incident but his tyre was badly gashed, and however suspicious he might be of us, he was obviously glad when we offered to repair the damage. We decided while conversing in our native Irish to charge one shilling for our labours. E. S. objected to the charge and offered to pay 6d. After considerable bargaining we agreed to "split the difference" at 9d . . . We were enjoying the encounter with what to us was a most unique character, even had he not spoken with a strange accent and not been dressed in an attire (Lederhosen) which we never saw before. His mode of dismounting the bike was no less fascinating for, instead of throwing one foot back while the other rested on the pedals, he threw it forward over the handlebars so fast that the movement of his hands to make way for his foot was barely noticeable. We did not know who he was until we saw his picture in the *Irish Press* about a week later.

O'Kennedy later studied science in Dublin and became professor of physical chemistry at University College, Galway. After Schrödinger had returned to Austria, O'Kennedy once wrote to him about some problem suggested by *What is Life?* By a remarkable coincidence he put on the envelope only a 3d stamp (inland postage) instead of a 6d stamp (foreign postage). Erwin, who did not in any way connect him with the erstwhile tyre mender, wrote a rather sharp protest concerning the Austrian schilling he had to pay in postage due.

DIAS INSTALLED

On June 1, the bill to establish the Dublin Institute for Advanced Studies (DIAS) passed the Senate, and it was signed into law by President Douglas Hyde on June 19. Schrödinger began his tenure as a senior professor at a salary of £1,200 a year, and it was a relief to be back on a payroll again. It was quite a good salary for Dublin, about what a successful solicitor would make. About a quarter of his salary would go to income tax. He was worried that the Institute made no provision for a widow's pension. The vicissitudes of his academic life and the loss of his retirement pay from Berlin and Graz meant that he did not have any appreciable financial reserves. He tried to save some of the Nobel prize as a safety net for Anny.

The Institute was located in two interconnected eighteenth-century Georgian houses. Numbers 64 and 65 on the south side of Merrion Square. To Brian O'Nolan, the Georgian architecture of Dublin was "a remote faded poignancy of elegant proportions, minute delicacy of architectural detail balanced against the rather charmingly squalid native persons who sort of provide a contrapuntal device in the aesthetic apprehension of the whole."

Within ten minutes' walk from Merrion Square are the grounds and buildings of Trinity College, in the heart of the city life of Dublin. The administrative offices of the National University of Ireland were on the east side of the square and one of its constituent colleges, University College Dublin, had its medical school on the south of St. Stephen's Green, adjacent to the Square. The Royal Irish Academy occupies Northland House which is nearby. All these learned institutions as well as the National Library and Art Gallery lie within an area not so large as the campus of many universities. At that time there were about 1,400 students at Trinity and 1,200 at UCD. In July, both universities conferred honorary degrees on Schrödinger.

The Institute buildings had been used for government office and some remodeling was necessary before they were ready for occupancy. Schrödinger's office was a spacious room, lined with bookshelves. There was a large desk with a comfortable chair, at the back of which was a Georgian fireplace in which the coals glowed on chilly days; even during the Emergency, the Institute managed to keep most of its fires burning. Erwin wore glasses and young visitors were at first intimidated by the "ice-blue eyes glaring greatly magnified through the strong lenses" but they were soon reassured by his kindly manner. As he worked or talked, he usually smoked his pipe, enjoying the ritual of cleaning, filling, tamping and relighting. If a visitor was also a smoker, the room would sometimes become so filled with smoke that they would have to repair elsewhere to replenish their oxygen supplies.

DIAS PROGRESSES

Schrödinger was happy with his situation. As he wrote to Born: "The people in Dublin are good to talk to but overworked with lectures and exams particularly . . . [My position] is one of the most appropriate in the world. To be reinstated to absolute security (at least as regards yourself) at 53 by a foreign government, in my case fills you with – well, infinite gratitude towards that country."

Even amid wartime stresses and shortages, DIAS flourished, fulfilling the highest hopes of de Valera, and Dublin quickly became a world center for theoretical physics. Almost every year, special colloquia brought contingents of the best theoreticians from the United Kingdom, including refugees from the continent of Europe. Special lectures were given by visitors at other times. Discussions were lively and the exchange of ideas stimulating. Graduate work in physics in the Dublin universities took on a new spirit. In addition to the specialized lectures and seminars, the statutes of the Institute provided for an annual public lecture series to be given alternately at TCD and UCD. These became one of the highlights of the intellectual life of Dublin.

16. First meeting of the Governing Board, School of Theoretical Physics, Dublin Institute for Advanced Studies (21 November, 1940). From left: Erwin Schrödinger, A. J. McConnell, Arthur Conway, D. McGrianna (Registrar), Eamon de Valera, William McCrea, Msgr Patrick Browne, Francis Hackett

Schrödinger recognized that DIAS was generously funded by the taxpayers of a relatively poor country with many demands on its resources. Therefore he always tried to give more than a full measure of service, especially during the war years when it was impossible to vacation abroad. He received many letters from the public, some of them long pseudo-scientific documents about which his opinion was requested. Sometimes he passed them along to assistants, saying "Now you see what I have to put up with," but every letter was answered.

Schrödinger usually arrived late at the Institute but he also stayed late. He would come in at about 11:30 a.m. while they were having tea, perch upon the table, and talk about all kind of things, but mostly physics and philosophy, until it was time for lunch. Paddy Browne was often there to debate with Erwin.

Anny and Hilde were also kept busy, for visitors had to be entertained. Buffet lunches and evening parties at Clontarf were quite frequent, and Anny became famous for her Viennese cakes and strudels. The Schrödingers were always kind and helpful to younger scholars and staff members, having them to tea or supper, while Erwin provided detailed advice on how to ride a bike safely in the wintry Dublin streets. He abhorred any formality of dress or pretentiousness of style.

Erwin preferred the company of young people as he grew older, especially young women, although he also was attracted to young men who were handsome, lively and intelligent. The Austrian community in Dublin used to meet at the "Old Vienna Club," located about a block from St. Stephen's Green. Soon after he came to Dublin Erwin met a young biochemist there, Stephen Feric, and they became good friends despite the twenty-five-year difference in their ages. Feric was full of youthful vigor and good cheer, and for a while he became practically a member of the Clontarf household. The housekeeping was divided on a rota system, with one week Anny then one week Hilde in charge. It usually worked smoothly, but occasionally there were tempests, and Erwin would retire upstairs until all was calm again. Sometimes, however, he and the two women would get involved in frantic scenes over trivial domestic problems, and Stephen would take Ruth into the garden or for a walk by the seashore.

Another young Austrian refugee encountered at the Old Vienna Club was Alfred Schulhoff, whose mother had gone to school with Hilde. He was practically adopted by the Clontarf family and they paid all the expenses for his study of electrical engineering at the university. When he thanked Erwin, the response was only, "Well, I might have had a son."

DUBLIN THEATER

By the time the Schrödingers arrived in Dublin, the great days of the Abbey Theater, of Willy Yeats, Augusta Gregory, Sean O'Casey, and John Synge, were only memories. The old theater still stood, however, at the corner of Marlborough and

Lower Abbey Streets, and its ten to twelve productions a year, except for those in the Irish language, often played to packed houses. The great rival of the Abbey, the Gate Theatre, was founded twenty-four years later, in 1928, by MacLiammoir and Edwards. The "boys" continued their partnership for almost fifty years, as actors, directors, designers, and lovers, but not always everything in every year. They did not neglect Irish drama, but their policy was to stage in Dublin the best plays from all countries and all times. They were enthusiastic about stagecraft and the Gate's sets were a relief from the whitewashed cottage walls and dismal city flats of the more realistic Abbey. As Micheal put it, "At that time the theatrical fare of Dublin consisted of an honest and supremely well cooked bacon and cabbage at the Abbey, stewed prunes and custard at the Gaiety, and kickshaws at the music halls."

Devoted to the theater since youth, Erwin soon came to know a lot of people in theatrical and artistic circles in Dublin, but he was never a close friend of "the boys" or their leading ladies Meriel Moore and Coralie Carmichael.

A gathering place of the bohemian set, where Erwin sometimes appeared, without Anny or Hilde, was Des McNamara's flat on the top floor of the Monument Café on Grafton Street, where Mac ran a non-stop Fabian salon in a one-room studio. He was noted for his papier-mâché theater masks. Erwin described himself to these friends as a "naive physicist," whose hobby was weaving tapestries. John Ryan had a studio next door, where Brendan Behan and the poet Paddy Kavanagh used to take refuge when the worse for drink, and Erwin would certainly have met Paddy but probably not Brendan.

Soon after their arrival in Dublin, Erwin, Anny, and Hilde were attending a play at the Abbey, and even Erwin had difficulty following the country brogue in which it was spoken. He turned to an attractive young man sitting next to him and asked, "What did he say?" This was Ronald Anderson, a school teacher and part-time actor. His wit and gaiety appealed at once to the Schrödingers and Ron became a good friend of both the ladies and also of Erwin.

Shortly before Christmas, 1940, David and Sheila (May)

Greene returned from Glasgow, where he had been lecturing in Celtic studies. He was appointed assistant librarian at the National Library, a post he held until he joined the Dublin Institute for Advanced Studies in 1948. Sheila came from an artistic Donnybrook family, who owned a large Dublin music store; her brother Frederick was one of Ireland's best known composers. In 1936, she made her debut with the Gate Theatre in Elmer Rice's *Not For Children*, and after returning from a successful tour to Egypt, became engaged to David in August 1938. Micheal called her "our new golden-haired *ingénue*"; she had rather unruly fair hair worn high off the forehead, large sympathetic blue-grey eyes set wide apart, and an air of serious animation. "She was kind and thoughtful and passionate in all that she did." Her good looks and charm caused "the better half of Dublin to be in love with her," in the words of Alexander Lieven, another young friend of Erwin's. Sheila's voice was not commanding enough for leading roles, so that she never became famous as an actress.

David Greene had a brilliant academic and social career as an undergraduate at Trinity College, taking double firsts in Celtic and Modern Languages, winning three gold medals, and captaining the water polo team. He was a big, handsome man, with an excellent brain but an irascible and self-centered disposition. Erwin was attracted not only by Sheila's beauty, but also by David's scholarship, and at that time he was making a serious, but ultimately unsuccessful effort to learn the Irish language. David considered that Erwin was "a very Socrates, although the Irish Catholic city fathers are fortunately too blinkered to prescribe hemlock." Erwin thought that David was "an amazing charming rascal."

TURN OF THE TIDE

Throughout the spring of 1941, the war continued to go badly for the Allies, as the Germans invaded and soon overran Yugoslavia and Greece. On April 15, an air-raid on Belfast left 500 dead. Schrödinger was worried and depressed, and he wrote to Weyl in Princeton:

Dear Peter:

Thanks for your gifts to the embryonic library of the Institute and greetings between institutes. To create anything that would deserve this name is in the present moment almost impossible. Continental books are well-nigh inaccessible. My copies of Courant–Hilbert i & ii, and of Jahnke–Emde, which I was fortunate enough to procure in France, will have to be the only ones for some time.

I should very much like to know how your people feel about the present situation . . . If you asked me my opinion, I should be at a loss to describe it properly. I have not a bit of hope. I have passed the stage of "désespérance." What happens is thus, that democracy has not been people's power, as the name indicates. It has been power of Eaton [sic] and Christ Church, and I know not what. Over here. Over there it has been Wall Street. May be all that has changed now. It was too late. Unless something spectacular comes forth from these democracies presently, I cannot see what should change the present state. I have believed long enough. I can no longer. It is all nothing. It is all old and weak and arteriosclerotic and inefficient.

On the night of May 31, the most serious bombing of Dublin occurred, with thirty people killed and many houses destroyed. The German government apologized for the "mistake," but most people regarded it as a warning. However, the course of the war was about to change.

Till now, Hitler had met no serious resistance in Europe, and he decided to complete his conquests by an attack on Russia, launching a massive invasion on June 22. That night Erwin noted in his diary: "The banner is turned around, the Proletariat of the world now knows on which side it stands . . . Considered quite soberly, it is really a great joy to see the two wretches [Hitler and Stalin] in battle against each other." For many months the issue hung in the balance, but ultimately the Red Army beat back the Nazi onslaught. The German military realized that defeat was inevitable and began to hatch various ineffective plots against Hitler.

From July 16 to 29, the first major seminar was held at the Institute, to coincide with the arrival of Walter Heitler as a new professor. Schrödinger gave a course of ten lectures on perturbation theory and the mathematical background for the quantum theory of the meson, and Heitler presented ten lectures on the meson theory of nuclear forces. Dublin had never before experienced such a deep and comprehensive treatment of an exciting topic in theoretical physics. Participants in the seminar came from all over Ireland. Paul Ewald came down from Belfast, where he now held a professorship. It was reported in the *Irish Times* as a conference on "The Secret of the Mystifying Mesotron."

Erwin wrote to Max Born to say how delighted he was to have Heitler at the Institute. "Scientifically he is the equal of his 'milk brother' London [they were both Sommerfeld students] but as a man and a teacher he is much greater in every respect. Reading his book has led me back to your nonlinear field theory." The book was *Quantum Theory of Radiation* a pioneering exposition of this subject. The book marked the directions in which future great advances in quantum theory would be made, and it is strange that even with Heitler as a colleague, Schrödinger was soon to turn away from quantum mechanics to pursue his researches in unified field theory.

THE MYLES CASE

In 1942, after he had been in Dublin about two years, Erwin had a comic theological collision with one of Dublin's most famous literary figures. This was Brian O'Nolan, an author of great wit and style, whose novels *At Swim Two Birds* and *The Third Policeman* are considered by some critics to be in a class with those of Joyce and Beckett. O'Nolan was a hard-drinking civil servant who wrote under a variety of pseudonyms. He wrote a daily column for the *Irish Times* under the title, *Cruiskeen Lawn*, "The Little Overflowing Jug," which he signed "Myles na Gopaleen," "Myles of the Little Ponies," after a character in a play by Boucicault. The local color and humor his material often concealed a savage social commentary.

Jack White, the feature editor of the *Irish Times*, was aware that Myles "took an extravagant pride in his ability to circumvent the laws of libel . . . His copy was scrutinized for scurrility and double meanings, and any column that offended was chopped ruthlessly or thrown into the wastebasket." It should be remembered that in the British tradition the laws of libel were designed to prevent the press from embarrassing the ruling classes, and material that would hardly raise an eyebrow in Boston could trigger legal salvos in Dublin. On April 10, 1942, *Cruiskeen Lawn* included the following paragraphs:

That nothing but the Best is good enough for the Institute of Advanced Studies is another quip that must, in the name of reticence and that delicacy of manner which distinguishes the gentleman, remain unsaid. Here, at any rate. By all means pass it off as your own in the boozer tonight. The laugh that you will get will be as forced and as false as your own claim to be a wit.

Talking of this notorious Institute (Lord, what would I give for a chair in it with me thousand good-lookin' pounds a year for doing "work" that most people regard as an interesting recreation), talking of it, anyway, a friend has drawn my attention to Professor O'Rahilly's recent address on "Paladius and Patrick." I understand also that Professor Schroedinger has been proving lately that you cannot establish a first cause. The first fruit of the Institute, therefore, has been an effort to show that there are two Saint Patricks and no God. The propagation of heresy and unbelief has nothing to do with polite learning, and unless we are careful this Institute of ours will make us the laughing stock of the world.

Schrödinger dismissed the column as a matter of no importance, but nevertheless a tempest of discussions, meetings, and telephone calls raged for the next ten days, until the meeting of the Council of the Institute on April 23. Solicitors were consulted and the editor of the *Irish Times*, R. M. Smyllie, was accused of publishing defamatory material. Schrödinger wrote to the Council the day before its meeting: "The draft of an apology by Mr Smiley [sic], Chief Editor, which I was shown yesterday by the Registrar and which was intended for publication in the *Irish Times*, includes the statement that I felt grieved by the article in question, and included an apology to me personally . . . I beg to decline emphatically the inclusion of any statement about my having been grieved by that article, or of any apology <u>to me</u>." The paper finally paid £50 to the Red Cross to settle the matter.

A full scale debate between Erwin and Myles would have

been a memorable occasion, for they both were masters of language and devotees of drama. They remained friendly and Myles consulted Erwin for advice about his adaptation of *The Insect Play* by the Čapek brothers, which was staged by the Gate Theatre early in 1943.

DIAS COLLOQUIUM

During the spring of 1942, plans were made for a major event in the School of Theoretical Physics, the first Colloquium with speakers from overseas. Dirac and Eddington agreed to come. The Colloquium met from July 17 to 26, with forty-five participants. Eddington's lectures were on "The Combination of Relativity Theory and Quantum Theory." These lectures were masterpieces of exposition, using hardly any mathematics; everybody could follow them without difficulty yet nobody could understand them. Dirac's lectures were on "Quantum Electrodynamics." He began by saying that his lectures, like those of Eddington, were concerned with unifying relativity and quantum theory, but his approach was first to find a neat and beautiful mathematical scheme, and then to fit it to a physical interpretation.

Schrödinger reported to Born that the Colloquium was

a great success <u>externally</u>: the two dear men here for a fortnight – pleasure, contact, rejoicing on all sides, receptions, parties, climaxed by huge invitation cards to meet the President on Wednesday afternoon . . . But the internal effect was negligible. Dirac gave a lucidly clear exposition of an intrinsically inconceivable business. Eddington entirely messes up a position which in its main points really is quite clear and tenable . . . As a matter of fact, they are monomaniacs. Your idea of getting their opinion on Born's theory is pathetic. That is a thing beyond their linear thoughts. All is linear, linear, – linear in the n'th power I would say, if that was not a contradiction. Some great prophet may come . . . "If everything were linear, nothing would influence nothing," said Einstein once to me.

Myles na Gopaleen took note of Eddington's remark that less than a thousand people in the world can understand Einstein's theory and less than a hundred can discuss it intelligently. He proposed to make it compulsory in the schools and to have it taught in Irish. Then instead of "being illiterate in two languages," Irish children could be "illiterate in four dimensions."

As soon after the Colloquium as possible, Erwin and Anny took off for a holiday in Killarney, County Kerry. They met a cheerful and friendly teenager, Lena Lean, whom they invited to come to Dublin to help take care of Ruth.

NON-LINEAR OPTICS

Schrödinger's principal research from 1941 to 1942 was devoted to a non-linear classical electromagnetic theory that had been devised by Max Born in 1934. He continued to explore the consequences of this theory throughout 1943–44, and some of the methods were then taken over into the more general unified field theories that became his main scientific interest until he left Dublin in 1956.

Schrödinger's motivation in the work may have been the intrinsic mathematical interest of the analysis, for the results he obtained led to no new conclusions that could be tested by experiment. Paul Dirac once said:

Of all the physicists I met, I think Schrödinger was the one that I felt to be most closely similar to myself . . . I believe the reason for this is that Schrödinger and I both had a very strong appreciation of mathematical beauty and this dominated all our work. It was a sort of act of faith with us that any equations which describe the fundamental laws of Nature must have great mathematical beauty in them. It was a very profitable religion to hold and can be considered as the basis of much of our success.

Whatever his motivation, Schrödinger devoted considerable effort from 1941 to 1943 to working out the detailed consequences of the Born–Infeld electrodynamics. His work had formal elegance and displayed his mathematical abilities, but it lacked any new physical ideas, being like a virtuoso variation on a classical theme. Nevertheless, his conviction that non-linear theories would be essential for future progress in physics has turned out to be justified.

17. Dublin Colloquium, 1943, Paul Ewald, Max Born, Walter Heitler and Schrödinger

UNIFIED FIELD THEORY

From 1943 to 1951, Schrödinger's research was dedicated almost exclusively to the search for a unified field theory that would encompass both gravitation and electromagnetism. He was inspired by a metaphysical belief in the unity of nature, a belief that had not essentially changed since he had written it down almost twenty years earlier in *Meine Weltansicht*. As he said then, only metaphysics can inspire the hard work of theoretical physics. The philosophy of Einstein in his later years was similar, leading to a feeling of wonder at the simplicity and beauty of the landscape of the universe as seen through the window of mathematical theory. Bruno Bertotti, who worked with Schrödinger from 1953 to 1955, called this view of the world "rational mysticism." Roland Barthes, however, discerned in it the perennial gnostic themes: "the unity of nature, the ideal possibility of a fundamental reduction of the world . . . the age old struggle between a secret and its utterance, the ideal that total knowledge can only be discovered all at once, like a lock that suddenly opens after a thousand unsuccessful attempts." Schrödinger was more a gnostic than a mystic. He never displayed any inclination toward the life of asceticism and self-denial that mark the way of a religious mystic. He would attempt to reach the secret at the heart of the world through a labyrinth of mathematical symbols.

Einstein was thirty-six years old when he published his general theory of relativity; Schrödinger was thirty-eight when he discovered wave mechanics. By 1940, when he began to consider generalized field theory, he was fifty-three. Einstein was then sixty-one and he had been working on the problem for twenty-five years without apparent success. In the history of physics it is unusual to find anyone who has made a major theoretical discovery after the age of forty. Thus it might seem that Einstein and Schrödinger were facing insurmountable psychological barriers. Revolutionaries in physics must be young men whose minds have not had time to become habituated to well-worn pathways of thought. In fact, however, Einstein and Schrödinger were not seeking any revolution, they were simply trying to extend the range of a method that had already proved its worth in the theory of general relativity. The techniques they used were well established in the 1916 papers of Einstein and Hilbert, the 1918 book of Weyl, and the 1923 book of Eddington.

At any point in history there are different potential directions for the future of physics, and the choice of a particular direction may preclude exploration of any of the others. When the choice is made, the best minds, a few in each generation, dash eagerly along the new path, and physics follows them, while the other paths are blocked by an impassable psychic gate. Thus Schrödinger and Einstein may have had little chance of success in their efforts to discover a unified field theory in the 1940s by 1920 methods. One can see an example of such a blocking of pathways in Richard Feynman's 1962 Caltech lectures on gravitation: "None of these unified field theories had been successful . . . Most of them are mathematical games, invented by mathematically minded people who had very little knowledge of physics and most of them are not understandable." Is it likely that any of his students would choose to work in such a direction?

It has been suggested that Schrödinger and Einstein turned to unified field theory because of their disenchantment with the prevailing state of quantum mechanics. Einstein, however, certainly hoped that field theory would eventually include both the macroscopic systems of cosmology and the microscopic physics of elementary particles and quanta. Schrödinger was less optimistic than Einstein: "At the back of our striving for a unitary field theory, the great problem awaits us of bringing it into line with quantum theory. This point is still covered with a deep fog."

There is no convincing evidence that either scientist turned to field theory as an escape from the uncertainties and probabilities of quantum mechanics. Einstein was continuing to follow the way that had led to his greatest success. Schrödinger, influenced by Einstein, had always been interested in the geometrization of physics, and the concept of a unified field was a counterpart of the unity of mind and nature in which he found ultimate reality. "The city of Brahman and the palace within it is the small lotus of the heart, heaven and earth are contained in it."

Schrödinger's first paper on unified field theory, "The General Unitary Theory of Physical Fields," was read at a meeting of the Academy in January 1943. The task he set for himself was to modify Einstein's general theory of relativity so that it would explain electromagnetism as well as gravitation. It soon became apparent, from this work and similar studies by Einstein, that electromagnetism can be "geometrized" in a bewildering variety of ways. The real problem is to establish a connection with the observable world, first by representation of the known physical laws, and then by prediction of hitherto unobserved relations. Einstein's general relativity had passed these tests brilliantly, but so far all the geometric unified-field theories had failed to do so. Before Schrödinger addressed this problem, however, he temporarily put aside mathematical physics to devote his mind to a problem that had fascinated him since student days – the mechanisms of heredity and the nature of life.

WHAT IS LIFE?

Schrödinger had agreed to give the statutory public lectures for 1943 and they were scheduled for February at Trinity College. He decided to prepare a semi-popular lecture on a biological subject, the mutation rate caused by action of X-rays on the fruit fly, *Drosophila Melanogaster*. He knew that the absorption of one quantum of X-radiation can produce a mutation at one locus in a chromosome. Perhaps he recalled this result from discussions with Max Delbrück in 1933 in Berlin. Sometime towards the end of September, he realized that if he knew the absolute intensity of the absorbed X-rays and the mutation rate they caused, he could calculate from the target area the effective size of the information center in the chromosome, the gene.

He must have mentioned this problem to Paul Ewald, for Paul sent him a copy of the paper from the 1935 *Göttinger Nachrichten* by Delbrück, Timofeef-Ressovsky, and Zimmer called "the green pamphlet" from the color of its reprint covers. This work had anticipated his estimation of the target area and suggested for the first time that a mutation is caused by a change at one place in a *molecule*. He began to think about the implications of this result for the physical and chemical mechanism of inheritance. He decided to give a series of three lectures under the title "What is Life?"

Among all the great physicists, why was Schrödinger the one destined to make a major contribution to the history of biological thought? The answer may be found in his early life: his father, the spiritual guide of his formative years, was passionately interested in biology, and Erwin always referred to him as a botanist, even though he was actually an industrial chemist whose botany was an avocation pursued in the time he could spare from his business. At the university, Erwin's intimate friendship with Franz Frimmel, the religious student of biology, reinforced his interest in this subject and led him into extensive reading in fields that were foreign to most physicists. Schrödinger was more than an amateur in biology; although he had not worked in genetics, he was still a world authority on the physiology and biophysics of color vision.

Erwin celebrated the New Year 1943 with Hilde and Ruth, while Anny was once again in Belfast visiting the Ewalds. The first lecture took place on Friday evening, February 5, with two more on successive Fridays. De Valera was there with other notables from Church and State. The number of people who tried to crowd into the Trinity lecture hall was so great that it was necessary to repeat the lectures on Mondays. The total audience was estimated as more than 400, and the numbers did not dwindle between the first and last lectures.

Even *Time* magazine took notice of the excitement in Dublin and published a story in its issue of April 5. "Schrödinger has a way with him," they wrote. "His soft, cheerful speech, his whimsical smile are engaging. And Dubliners are proud to have a Nobel prize winner living among them. But what especially appeals to the Irish is Schrödinger's study of Gaelic, Irish music and Celtic design, his hobby of making tiny dollhouse furniture with textiles woven on a midget Irish loom – and, above all, his preference for a professorship at the Dublin Advanced Studies Institute for one at Oxford."

As soon as the lectures were finished, he began to prepare them for publication as "a little booklet." The basic question was clearly stated at the beginning: How can the events *in space and time* which take place within the boundary of a living organism be accounted for by physics and chemistry? At present, he said, these sciences cannot answer the question, but in the future they will be able to do so. The reason no answer can be given now is that the most essential part of the living cell, the chromosome fibre, is a piece of matter that differs entirely from any matter hitherto studied by the physicist. It may suitably be called an *aperiodic crystal*. By this is meant a regular array of repeating units in which the individual units are not all the same. "Incredibly small groups of atoms, much too small to display exact statistical laws, do play a dominating role in the very orderly and lawful events within a living organism."

Schrödinger next gave an overview of the current understanding of the mechanism of heredity, in terms of chromosomes, genes, mitosis, meiosis, and crossing over. He introduced here what was to become one of the most fundamental concepts in the new science of molecular biology: *the chromosome is a message written in code*.

In calling the chromosome fibre a code-script we mean that its structure determines whether the egg would develop, under suitable conditions, into a black cock or into a speckled hen, into a fly or a maize plant, a rhododendron, a beetle, a mouse or a woman . . . But the term code-script is, of course, too narrow. The chromosome structures are at the same time instrumental in bringing about the development they foreshadow. They are law-code and executive power – or to use another simile, they are architect's plan and builder's craft – in one.

This was the birth of the concept of a *genetic code*. A few earlier works had hinted at the idea of a code in the chromosomes, but Schrödinger was the first to state the concept in clear physical terms.

The chromosome fibre is a linear code for the information that constitutes the genotype of a particular individual organism. The smallest locus in the chromosome coding for an individual difference in the genotype is called the *gene*. Schrödinger defines it as "the hypothetical material carrier of a definite hereditary feature."

What is the maximum size of a gene? Crossing-over data and microscopic examination of giant salivary-gland chromosomes from *Drosophila* roughly agreed that at most a gene can contain a few million atoms. He thought that the gene is probably a large protein molecule. Almost at the same time that these words were being spoken, however, Oswald Avery, then sixty-five years old and facing retirement from the Rockefeller Institute for Medical Research in New York, was writing up his most recent experiments which showed that the "transforming principle" in *Pneumococcus* is not a protein but is in the DNA (deoxyribonucleic acid) fraction from the cell nuclei.

The permanence of genetic information is almost absolute.

"The whole (four dimensional) pattern of the 'phenotype,' the visible and manifest nature of the individual . . . is reproduced without appreciable change for generations." When a change does appear it is discontinuous or "jump-like," occurring in a few individuals out of a large population, and there are no intermediate forms between the few changed and the many unchanged. Such a discontinuous change is called a *mutation*. "It reminds a physicist of quantum theory – no intermediate energies occurring between two neighboring energy levels."

The existence of stable chemical bonds between atoms in molecules was explained only after the advent of quantum mechanics. Heitler and London, working in Schrödinger's department in Zurich, devised the first quantum mechanical theory of the chemical bond in 1927. The development of ideas may thus be summarized as follows: discontinuity in physics→quanta of energy→quantum theory of chemical bond→stable molecules→genetic code→discontinuity in genetic mutations.

Schrödinger explained that the nature of the chemical bond is essentially the same in molecules and crystals; thus it is correct to call the gene molecule an aperiodic crystal, a long linear molecule in which the units are not identical but consist of a small number of moieties that can serve as the symbols in the code script, like the dots and dashes in the Morse code. As an example he considered a code made up of twenty-five units consisting of five each of five different symbols. He calculated that there are 62×10^{12} possible combinations. If Schrödinger had been aware of the recent work on *Neurospora crassa*, the bread mold, by George Beadle and Edward Tatum at Stanford University, he might have gone on to surmise that genes were coding for protein structures, but his knowledge of genetic research was not sufficiently up to date. Nevertheless, he indicated that "the miniature code should be in one-to-one correspondence with a highly complicated and specified plan of development and should somehow contain the means of putting it into operation." How does it do it? Schrödinger did not expect any detailed information on this question to come from physics in the near future, but he was confident that biochemistry guided by physiology and genetics would eventually provide an answer.

Schrödinger then made a statement that, when the lectures were published, inspired many young physicists to turn to research in biology. "From Delbrück's general picture of the hereditary substance it emerges that living matter, while not eluding the 'laws of physics' as established up to date, is likely to involve 'other laws of physics' hitherto unknown, which, however, once they have been revealed, will form just as integral a part of this science."

He believed that these other laws of physics would be related to the ability of a living organism to maintain itself in an ordered state and to evade the tendency toward equilibrium of all non-living systems. "How does a living organism avoid decay? The obvious answer is: by eating, drinking, breathing . . . What then is that precious something contained in [its] food which keeps [it] from death? . . . It can only keep . . . alive by continually drawing from its environment negative entropy." Schrödinger then fell into a partial error, which was later pointed out to him by Franz Simon and others. He seemed to identify the source of negative entropy as the orderliness of the molecules that are eaten: "An organism maintains itself stationary at a fairly high level of orderliness (= fairly low level of entropy) [by] continually sucking orderliness from its environment." This statement might be misleading since it emphasizes the inward flux and neglects the outward one. The organism is an open system; its entropy can decrease or remain constant provided the total entropy of organism plus environment always increases; thus it is not merely a matter of ingesting organized foodstuffs, rather the net overall flux of entropy from organism to environment must be positive.

The new kind of physical law that Schrödinger discovers in life is rather an anti-climax when he finally propounds it: the living organism is like a clock and life is like clockwork.

Thus it would appear that the "new principle," the order-from-order principle, to which we have pointed with great solemnity as being the real clue to the understanding of life, is not at all new to physics . . . We seem to arrive at the ridiculous conclusion that the clue to the understanding of life is that it is based on a pure mechanism . . . The conclusion is not ridiculous and is, in my opinion, not entirely wrong, but it has to be taken with "a very big grain of salt."

The cardinal point is that clocks are made of solids, which are kept in shape by the Heitler–London forces that make chemical bonds. The living organism depends on an aperiodic crystal, the chromosome fibre. One may call it a cog of the organic machine, but this cog "is not of coarse human make, but is the finest masterpiece ever achieved along the lines of the Lord's quantum mechanics."

On this pious note, Schrödinger ended the lectures at Trinity College, but when he prepared the written version for publication he added an epilogue, "On Determinism and Free Will":

Immediate experiences in themselves, however various and disparate they be, are logically incapable of contradicting each other. So let us see whether we cannot draw the correct, non-contradictory conclusion from the following two premises:

(i) My body functions as a pure mechanism according to the Laws of Nature.
(ii) Yet I know, by incontrovertible direct experience, that I am directing its motion, of which I foresee the effects . . ."

The only possible inference from these two facts is, I think, that I – that is to say, every conscious mind that has ever said or felt 'I' – am the person, if any, who controls the "motion of the atoms" according to the Laws of Nature. In Christian terminology to say "Hence I am God Almighty" sounds both blasphemous and lunatic. But please disregard these connotations for the moment and consider whether the above inference is not the closest a biologist can get to proving God and immortality in one stroke.

Schrödinger then relates this idea to the expression in the earliest Upanishads: Atman = Brahman, the personal self is identical with the all-comprehending universal self. Far from being blasphemous, this idea, the essence of Vedanta, is to him the grandest of all thoughts.

He continues in a more personal vein. In the West, such

unity may at times be experienced by true lovers who, "as they look into each other's eyes, become aware that their thought and their joy are *numerically* one – not merely similar or identical; but they, as a rule, are emotionally too busy to indulge in clear thinking, in which respect they very much resemble the mystic." Erwin was not writing now on the basis of pure theory – he was falling in love again, at the beginning of what was to be the last major love affair of his life.

Much of what he says in the epilogue repeats what he wrote in *Meine Weltansicht* eighteen years previously. Now, however, he lashed out savagely at "official western creeds," accusing them of "gross superstition" in their belief in individual souls. Plurality of selves is merely an illusion produced by Maya, like a deception seen in a gallery of mirrors.

WHAT IS LIFE? – THE BOOK

In view of his situation in Ireland, it is astonishing that Erwin at this time would publish a scornful denunciation of western religious beliefs, which was certain to dismay many of his colleagues and friends. His tone would have been appropriate for a Voltaire or Diderot in eighteenth-century France, but it was strangely discordant with the spirit of tolerance that prevailed, at least in intellectual circles, in contemporary Dublin. He was then fifty-six years old, an age by which wisdom and prudence might have been expected to temper polemic with mildness and courtesy.

He made arrangements to publish *What is Life?*, including the controversial additional chapter, with Cahill & Co., a respected Dublin publisher. The book was in the final stages of publication, when someone raised a question about the controversial sections. Possibly this was Paddy Browne – he was helping Erwin to polish his English and when he came to the epilogue, he was furious, especially since his help was acknowledged in the preface. Thus not only did Schrödinger propose to publish derogatory remarks about religion, he was even thanking a monsignor at Maynooth, the training center for Irish priests, for his co-operation.

The situation here recalls the letter that Schrödinger wrote in Graz at the time of the *Anschluss*, a letter which he refused to disown even when he knew that it would hurt his friends among the refugee scientists. "What I have written, I have written," he said at that time in Oxford, and he probably used similar words in Dublin. The consequence was that Cahill & Co. refused to publish the book and the type was dispersed. Erwin and Paddy Browne remained friends, but they never again achieved the warm and trusting relationship that had marked their earlier years at the Institute.

In October, Schrödinger sent a copy of the manuscript to a friend in London, the physical chemist Frederick Donnan, who recommended the Cambridge University Press as a suitable publisher. The book finally appeared late in 1944.

Hermann Muller gave *What is Life?* a favorable review in *The Journal of Heredity*. He dealt gently with the fact that Schrödinger failed to mention his priority in the discovery of X-ray mutations. Muller was appalled, however, by the mysticism of the final chapter: "If the collaboration of the physicist in the attack on biological questions finally leads to his conclusion that 'I am God Almighty,' and that the ancient Hindus were on the right track after all, his help should become suspect." Such a distrust of the mysticism of modern physics was to be echoed many times by biologists. With few exceptions, notably Charles Sherrington and his pupil John Eccles, they remained firmly attached to the world view of nineteenth-century physics, to mechanism, reductionism, and materialism. They experienced no metaphysical torments about ultimate reality, since they had found it in physics itself, an essentially classical, deterministic physics.

Schrödinger's book had an enormous influence, not only upon physicists who were persuaded that their methods might solve the problems of biology, but also among biologists who were encouraged to think more rigorously in terms of mathematically formulated and physically testable models. It was translated into seven languages and the total sales are estimated as over 100,000 copies. Erwin did not derive much profit from such sales; for example, a printing of 10,000 of the Japanese edition netted him £17.

In the autumn of 1946, the National Academy of Sciences sponsored a conference in Washington on "borderline problems in physics and biology." Max Delbrück opened the discussion by saying that Schrödinger's book had caused them to come together for what would doubtless be the first of many such meetings.

What is Life? had a determining influence on the career of James Watson. He read the book in the spring of 1946, while he was an undergraduate at Chicago and undecided what to do and where to go for graduate work. "From the moment I read Schrödinger's *What is Life?* I became polarized toward finding out the secret of the gene."

Francis Crick thought that the book was "peculiarly influential" and "attracted people who might otherwise not have entered biology at all." Maurice Wilkins reported that "Schrödinger's book had a very positive effect on me and got me, for the first time, interested in biological problems." Seymour Benzer was fascinated by Schrödinger's concept of the "aperiodic crystal."

No doubt molecular biology would have developed without *What is Life?*, but it would have been at a slower pace, and without some of its brightest stars. There is no other instance in the history of science in which a short semi-popular book catalyzed the future development of a great field of research. The influence of the book continues to be felt, and many people who know nothing else about Schrödinger will immediately associate his name with *What is Life?*

SHEILA

On March 15, 1943, Sean O'Casey's new play *Red Roses for Me* opened at the Olympia Theatre, and Sheila May (Greene) played the important role of Sheila Moorneen, the Catholic lass in love with the young Protestant leader of the railway strikers. By this time Sheila had been married almost five years. There were no children and David at least was determined never to have any, thus she had decided to resume her career as an actress and to write a master's thesis in social work at Dublin University. The part in *Red Roses* was her best so far. As

18. Sheila May Greene, from the painting by Barbara Robertson

an active member of the Irish Labor Party, her political sympathies were all with O'Casey. Sheila May published a critique of the play in the *Dublin Magazine*. She objected to the character she had played:

Sheila is the latest addition to that dreary procession of Mary Boyles and Norah Clitheroes and Iris Ryans, weak and clinging, hugging security rather than her lover, with his dreams and aspirations, losing him to clammy death and realizing too late, etc. At least she wasn't wronged – we are spared that. Why is it that neither Sean O'Casey nor Paul Vincent Carroll can draw a decent, sensible, full-blooded young woman? There must be a few Pegeen Mikes still left in Ireland.

Sheila did not hesitate to attack the social irresponsibility of the de Valera government. Since early 1942, Dublin had been stirred by a controversy over a plan by the City Council to provide a hot lunch for school children. The wrath of the bishops was aroused and they accused the Council of interfering with the normal family life of the people. The leaders of the Church, who watched every schoolgirl's knee with anxious eyes, had no concern for undernourishment and rickets. Sheila attacked de Valera directly in an article in *The Bell*, Sean O'Faiolain's literary monthly. She visited some of the worst Dublin slums and interviewed the mothers.

What do these people eat? I asked . . . The menu seldom varies: milk and bread and butter for the children, tea and bread and butter for the adults . . . Not one of these people, children or adults, ever touches fresh fruit or vegetables. Nearly all the children in this alley suffer from one complaint or another – TB, rickets, scabies and conjunctivitis are the most common. There is no place for them to play, except a passage in front of the houses about three feet wide, or the courtyard which houses the three lavatories for ninety people. The last time I visited the place the sewer pipe under the front archway had burst, and a harassed father was urging the children not to paddle in the sewage on the ground above it.

She cursed the "ignorance and inertia" which allowed such conditions. Her protests were disregarded, as de Valera did not dare to take any measures opposed by the bishops.

Meanwhile, for reasons that were not concerned with Erwin, Sheila and David began to have some serious quarrels. When angry, he could become quite violent, and she was rather afraid of him, but at the same time rather enjoyed the excitement. One time when they were visiting Dunquin, she hid herself in the bushes for fear he might throw her over the cliffs. Erwin and Sheila began to meet more often. They were not yet lovers, but his thoughts were turning to the composition of poetry as a change from tensor analysis.

In all his previous love affairs, except that with Hansi Bauer, Erwin had been the supreme male confident of his superiority over psychologically subservient women. "Poor things," he called them, "they have provided for my life's happiness and their own distress. Such is life." In Sheila May he found a woman who, except in science, was more nearly his equal. They were very different in their views of the world. She was a political activist and he was an ivory-tower intellectual.

In the Springtime of 1944 Sheila and Erwin became lovers. He wrote in his journal: "What Is Life? I asked in 1943. In 1944, Sheila May told me. Glory be to God!" At first they met at her house, and were involved in the cryptic messages, stolen glances, anxious phone calls, and hasty assignations that mark the typical adulterous affair. On July 24, Erwin rented a little apartment in the center of the city, near to both her house and the Institute. Now they could enjoy their love in a habitation of their own, but still, as Erwin wrote, "No word, no line, May reach the world, We quench our pride, And in ourselves abide."

Most of the love poems that Erwin published in 1956 were written to Sheila.

> *Liebeslied*
> Niemand als du und ich
> Wissen wie uns geschehn.
> Keiner hat es gesehen
> Wenn wir uns küssten inniglich.
> Keiner, keiner weiss
> dass uns der Himmel liebt
> dass er uns alles gibt
> was er zu geben weiss.
> Und säh uns wer
> er dacht es kaum

dass in weiten raum
sonst alles leer,
nur wir, nur wir
und unser glück
Nie nie zurück
als nur mit dir.

Lovesong
No one knows as you and I / How with us it came to be. / Not a one
was there to see / When we kissed so fervently. / No one, no one
knows / That heaven loves us so / That it gives us every thing / That
how to give it knows. / And whoever might see us / Would scarcely
think / That in wide space / Of all else void, / Are we alone, only we /
And all our joy. / Never never do I return / Except with thee.

In this love, Erwin almost found the mystical union
promised by the Hindu scriptures. In a passage from his
journal addressed to Sheila: 'You might give me what you
please in the future, and you probably will be pleased to give
me quite a lot, but nothing will ever outweigh the moment
when I saw the glory of God with all his angels, when your
half-open lips, quivering as it were (or was that me?) . . . told
me that you loved me." This recalls Francesca and Paolo,
"questi, che mai da me non fia diviso, la boca mi bacio tutto
tremante," or indeed Lancelot and Guinevere, and all the
romantic lovers of the age of chivalry. Two months later,
Erwin wrote that "with all that – ostensibly the ultimate
fulfillment of the supreme expression of love, yet there still
remains a longing for something more, something . more
intense, a more complete union. Something like blood flowing
into blood (she said) some more intimate unification."

Unless he went to elaborate lengths to delude posterity, it is
evident from his journals that Erwin was not, or not usually, a
libertine. He speaks the authentic language of romantic love,
seeking transcendence in the person of his beloved. In one
respect, however, he is not a romantic: he does not idealize the
person of the beloved, his highest praise is to consider her his
equal. "When you feel your own equal in the body of a
beautiful woman, just as ready to forget the world for you as

you for her – oh my good Lord – who can describe what happiness then. You can live it, now and again – you cannot speak it." Of course, he does speak of it, and almost always with religious imagery. Yet at this time he also wrote, "By the way, I never realized that to be nonbelieving, to be an atheist, was a thing to be proud of. It went without saying as it were." And in another place at about this same time: "Our creed is indeed a queer creed. You others, Christians (and similar people), consider our ethics much inferior, indeed abominable. There is that little difference. We adhere to ours in practice, you don't." Whatever problems they may have had in their love affair, the pangs of conscience were not among them.

Sheila was as much an unbeliever as Erwin, but in a less complex, more realistic way. She was never entirely convinced by his vedantic theology. She wrote to Erwin,

The difficulty is to me that what we call consciousness, all the experiences that build us into the creatures that we are, each with his own individual memory – these experiences come always through the senses . . . I know Erwin through sight, touch, and sound. If I cannot see, or feel, or hear him, does he exist for me at all? If I had not got these senses, how could I be aware of his existence, or indeed of my own? Take away these things by death, and what is left? Is there anything else? Ordinary common sense tells me no. Of course I want to believe differently, but then we might as well embrace the Christian myth or any other. Is it possible that your intelligence being too great to accept the usual fairy stories is forced to invent a more rational explanation of its own. Please, darling, don't think I have the temerity to criticize a brain so far superior to mine. I am asking, asking questions all the time, because I want so much to know and I must know. I feel now as if I were standing before a half-open door, and I know that I can only go through it hand in hand with you. Perhaps I shall be left behind.

One thing I do know is that I love you in a way that will last as long as I last, and that when we have kissed and made love all we want to, this other stronger force will live on – at least for me . . . I have never felt like this before.

Now do be a good man and destroy this letter, or lock it away, but don't leave it in a telephone booth. I love you very much . . . Sheila.

Erwin would summarize his sayings to Sheila for his journal. Early in August:

Sheila, please do not think I am a complicated man for whom his so-called "brain work" plays a prominent role in itself and is linked with his natural life by most involved, curved, branched channels. No. They are neighbours. Both are equally simple and straight-forward, equally natural. I have never been able to understand, let alone explain, anything difficult or mysterious or involved. I hate it. The simplest thing in the world is to go to bed. We have to do it every day. And everybody hates to do it alone. And you have given me more, more, a thousand times more than anybody ever has: your clear, clean, simple, straightforward love. Not for one second has there ever been any petty play about it, nor will there ever be.

Arthur Schopenhauer saw "falling in love" as the process whereby the noumenal, Kant's "thing in itself," enters the world of phenomena. He had more to say about sexual love than any previous philosopher, since "the sexual relation in the world of mankind . . . is really the invisible central point of all action and conduct . . . The ultimate aim of all love affairs, whether played in sock or in buskin, is actually more important than all other aims in man's life; and therefore it is quite worthy of the profound seriousness with which everyone pursues it. What is decided by it is nothing less than the *composition of the next generation.*" The new individual who will arise from the love affair is like a new Platonic idea, and "just as all the Ideas strive to enter into the phenomenal with the greatest vehemence, avidly seizing for this purpose the matter which the law of causality divides among them all, so does this particular Idea of a human individuality strive with the greatest eagerness and vehemence for its realization in the phenomenon. This eagerness and vehemence is identical with the passion for each other of the two future parents."

It is not likely that Erwin was considering Schopenhauer's *The Metaphysics of Sexual Love* during his meetings with Sheila. He protested that he was not trying to regain his youth: "I do not deplore anything I have lost, but quite often a thing I have kept. Considering my age, it is unlikely that any thing more precious will come my way . . . Goethe, Böcklin, Legantini, the old Franks and Provençals, the Upanishads – and you." And then "Gib mir, gib mir Ihr Kind." A few days later, Sheila's pregnancy was confirmed. "I am the happiest man in Dublin, probably in Ireland, probably in Europe. At least I cannot imagine greater happiness."

As is inevitable in human affairs, the mystical union of sexual love did not endure for long – with Erwin it was never able to survive tidings of pregnancy, but Sheila had hoped for something more. In the middle of August, she wrote:

I looked into your eyes and found all life there, that spirit which you said was no more you or me, but us, one mind, one being, one loving. For two months that common soul existed. Today I saw the scales creep over your eyes and I watched it die. It slipped away without even a struggle. My mind went numb, there was nothing I could do, or ever can do, to give us that again – You love me still, I know. I love my cat because he's soft and sweet and lets me play with him. You can love with tenderness, with devotion, you can love me all your life, but we are two now, not one. Why did you let it go? Wasn't it worth fighting for? My fault maybe in the beginning, for I am thoughtless and foolish, but surely age and learning bring some sort of wisdom to a man . . . Don't you know that anything could be achieved when you and I are together, that even though I am young and scatter-brained, when you open your mind to me, I can see with it and use it. But no, you talk of one lover placing the other on too high a pedestal. You talk of loving, but perhaps not approving. In a few brief sentences you kill the greatest thing I ever had, and then you ask me into bed, unless I would prefer to go out for a drink. Of course, in bed we're all right, we'll always do that well. But what is gone, will it ever, ever come again? I could stand deliberate cruelty from you and I wouldn't really mind, but the heart-breaking thing is that you didn't even know what you were doing. I can only pray that our child has been conceived a week ago, or two.

In the middle of September, another cloud appeared on the horizon of personal relations. Anny was again in Belfast with the Ewalds and she reported a serious misunderstanding with Paul and Ella. For the past five years Anny had lived as a loyal companion and housekeeper in the Schrödinger home, with only her visits to Belfast and her faithful dog Barney to provide any relief for her need for love and affection. There is no evidence that she and Paul were at any time lovers in a physical sense, but her love for him had become a psychological necessity. When the tension of this became too great for both of them, he tried to withdraw from the situation, and

Anny was devastated by the threatened loss. When she returned to Dublin she was close to a breakdown. Erwin said that he could not help her in dealing with her sweetheart, but he was nevertheless upset by this emotional crisis. An excursion made by Erwin and Sheila to Cork was unsatisfactory for both of them. She was worried about the baby and about what to tell David, who would have little reason to think that it was his. Erwin was so concerned about the situation of Anny and Paul that he seemed uninterested in Sheila's problems, just at the time she most needed his love and attention. On October 7, he wrote to advise her that the most prudent thing would be to go to David and tell him that the affair was ended. "Two months delight, and eight month's worries," he wrote in early November. But a few days later, he was more cheerful,

> She muses and jubilates and weeps and laughs
> In many a long winter night
> And dreams in joy and many pains
> Of my child under her heart.

In the middle of November, Sheila wrote to Erwin:

I think I told you that I believe that the love generated between two people is far greater than the persons themselves. Sooner or later the strain of living up beyond their powers will cause a fall and that is why so many marriages fail and so many people are unhappy and disillusioned . . . Having no supernatural God to worship I put you in his place . . . I want to finish my thesis . . . I'd like a nice hole in the ground where I could live as a rabbit . . . *P.S.* I have a lecture at 12 tomorrow so come at 10:30. I can send the maid out shopping.

As the old year came to an end, so did the brief but incandescent love affair of Erwin and Sheila. David behaved very decently about the whole thing. He knew that his wife wanted a child that he was unable or unwilling to give her. A few years later, Sheila and David separated, but David kept the child and brought her up as his own. In 1959, he married the sculptress Hilary Heron. She made a bronze nude of him in the style of Epstein's Adam; it was kept in the foyer of their house and much admired by visitors. He had a distinguished career as one of the foremost Celtic scholars of his generation.

Sheila became a political journalist, editing the newspaper of the Irish Labor Party for a while. Late nights and heavy drinking with male colleagues took their toll and she died in the early 1970s.

Erwin's last poem to Sheila was "Der Entäuschte" (The Disappointed One), which included the verses:

> Du bist nicht schuld. Du gabst mir alles gern
> worum ich bat. Gabst mir im überfluss.
> Trägst mir ein kind. Doch schlimmerts den verdruss
> dass du mir jetzt so tausend meilen fern.

> You are not to blame. You gladly gave me all / For which I asked. Gave me to overflowing. / Bore me a child. Yet that makes worse the discontent / That now you are to me a thousand miles away.

Post-war Dublin

After five years of the Second World War the end of the conflict
and the defeat of Germany and Japan could be foreshadowed,
so that the Irish "Emergency" was becoming less anxious.
Schrödinger no longer confined his work to unified field theory
and he began to take an interest in a variety of other problems,
including some of a more philosophical nature. Physically, as
he neared the age of sixty, he enjoyed good health, despite
annual attacks of respiratory illness, aggravated by the Irish
climate and excessive smoking. When springtime came, Dublin
broke out with great masses of yellow daffodils for sale at
almost every corner and as he strolled down O'Connell Street
he exclaimed in joy at the sight of all the pretty girls who had
emerged from winter coats to warm themselves in the occasional midday sunshine.

STATISTICS AGAIN

From January to March 1944, Schrödinger had returned to
one of his first loves in science in a course of lectures on
"Statistical Thermodynamics" at DIAS. They were published
(1946) by the Cambridge University Press. In less than a
hundred pages he covered the fundamentals of the subject with
an insight and clarity that have never been equaled. The book
is a distillation of his many years of creative work in the field,
and one hears echoes of the passionate discussions of the 1920s
with Planck, Ehrenfest, and Einstein. He was pleased to learn
that Max Born liked the little book and wrote to him: "For I
have no higher aim than to work out the beauty of science.

I put beauty before science. *Nitimur in vetitum* [Ovid: we strive for that which is forbidden]. We are always longing for our neighbour's housewife and for the perfection we are least likely to achieve."

During 1944 Schrödinger exchanged several letters with Lajos Janossy (1912–78), a Hungarian physicist who had attended some of his courses in Berlin, and who was now working on cosmic rays with Patrick Blackett in Manchester. He had become interested in a note by Janossy in *Nature* on the statistics of coincidences in Geiger counters – anything to do with statistical problems attracted his attention, and he would put aside even his general field theory to deal with it. Janossy was invited to lecture at a Dublin summer school on cosmic rays in 1945, and he stayed on to direct a new cosmic-ray section at the Institute. He was an intense young man with a shock of black hair that was continually falling into his eyes. A stepson of Gyorgy Lukács, he was also himself an ardent Marxist, but he was a great admirer of Erwin and they got along well together.

The group of young researchers at the Institute now consisted of James McConnell, H. W. Peng from Peking, Jim Hamilton, and Sheila Power. Peng noted a basic reason why Schrödinger had so few students: while he was thinking about a problem and trying to obtain a clear picture in his mind, he did not like to communicate with anyone, and as soon as it became clear to him, the result was so polished that there was nothing left for a student to do. Peng recalls that Schrödinger once illustrated a lecture with a plaster model which he had made of a potential surface. Later, after a dinner at Clontarf, Peng was examining an artistic sculpture of Venus, and was amazed when Anny told him that Erwin had made it himself.

During 1945 Schrödinger had an extended feud with Registrar McGrianna over the supervision of the cleaning ladies. When his office was tidied, notes that he had thrown onto the floor sometimes disappeared into the trash bins. A special meeting of the governing board was called to consider the problem, and when it failed to affirm the absolute authority of the director, Schrödinger handed in his resignation from that office. He refused a plea to reconsider, and Heitler was appointed director from the beginning of 1946.

TESTS OF UNIFIED THEORY

In 1944, Schrödinger published three papers on unified field theory. He enlisted Father McConnell in an attempt to find some experimental support for his predictions of an extra magnetic field around the earth or sun. McConnell made extensive analyses of the available geophysical and solar data, but the experimental uncertainties were too large to permit any unequivocal results.

Viewed in retrospect, the work on unified field theory does not appear to have led to important advances in physics, but one should not assess the value of scientific work merely in the light of subsequent results. In Schrödinger's own view at the time, it was as exciting and important as any work he had ever done: "I have found the unitary field equations. They are based only on primitive affine geometry, a way which Weyl opened and Eddington extended, whereupon Albert did the main job in 1923, but missed the goal by a hair's breadth. The result is fascinatingly beautiful. I could not sleep for a fortnight without dreaming of it."

Albert, however, after twenty-five years of frustrated pursuit of the same goal was less sanguine: "Concerning an affine solution of the electrical problem, I have become quite skeptical. This thing, alongside so many others, has been relegated to a pretty spot in the graveyard of my enthusiastic hopes – at the time I found it difficult to separate myself from it . . . One thing is certain. The Lord has not made it easy for us. As long as one is young, one does not notice this so much – luckily."

A NEW LOVE

It may be that Erwin was convinced from previous experience that scientific creativity would be promoted and sustained by erotic excitement, but even if he did not consciously decide to act on this principle, the fact is that he had hardly ceased making love with Sheila when he embarked upon a new amorous adventure.

Erwin was introduced to Kate Nolan (a pseudonym) by Hilde March, who met her through her work with the Red Cross. Kate was twenty-six years old, a young woman of limited education who worked in a government office in the city. She was a tall, slim, brown-eyed blonde, not so pretty as Sheila and completely lacking her sophistication, but having all the freshness and charm of a strictly brought-up Irish maiden. Her father had died when she was young and she had been raised by her mother and grandmother in accord with strict Catholic principles. Erwin admitted that it was not easy to convince himself that he was in love with Kate, but such a conviction was a necessary (and sufficient) condition for a seduction. Kate was at first rather shocked by the news that Erwin loved her, but he explained "If you can't love me as a lover because I am a married man, just love me as a father."

Erwin had never devoted much attention to the upbringing of his own daughter Ruth, who was now eleven years old and beginning to be more aware of the world of adults. At about this time she told him in the searching way that children have: "Ervinilly, you'll never be a grandpa." "Why not?" "Because you have no children." "Are you sure?" "I'd laugh if you had." "Well, laugh then!" Not until many years later did he tell Ruth that he was her father. She was about seventeen, and they were on a beach in bathing suits, when he said "Look at your feet – they are exactly like mine."

On May 8 came VE Day: the "Emergency" was over after almost six years. The event was marked by a riot at Trinity College. About 130 students made their way to the roof of a building facing College Green and on a flagpole ran up the Union Jack, the Stars and Stripes, and the Hammer and Sickle, in that order. At sight of the hated British flag all traffic in College Green came to a halt, and citizens rushed the college gates, assaulting students who tried to bar their way, and smashing everything that was breakable.

Thus it was in an atmosphere of general optimism that Erwin continued to woo Kate. In July he took her to Wicklow,

19. Erwin with his daughter Ruth (1946)

a quaint seaside town south of Dublin, and wrote for her a poem about their outing.

There must be something infinitely appealing for a naive young woman to have poetry written especially for herself by a love-inspired professor. Erwin did not hesitate to include even the deity in his verses, imploring God to give him Kate, "the only bliss to stay by me for ever."

While Kate was gradually falling under the spell of Erwin's charm, Sheila's pregnancy was progressing. Although they were no longer lovers, he often tried to encourage her when she became despondent about David's lack of interest in the forthcoming addition to his family. She told him, "If you cannot love me as a lover because I am already married, make up for it by loving me as a daughter." They often met in the mornings on the banks of the Liffey and talked at length until he had to rush away to meet Kate in the city during her lunch hour.

Sheila's baby, a girl, was born on June 9. Sheila was happy and adored the baby, and even David appeared to welcome it. As Sheila, never at a loss for words, remarked, "He has always

20. On the banks of the Liffey

been longing for someone to like him better than anyone else –
he has lost his wife and his dog, but now he has the baby."
Erwin saw Shaw's *Arms and the Man* that night, and commented
that "All great things in the world are worked through love –
not only children. It produces everything. Love is not an
impediment to great effort but its carrier." By that time he was
convinced that he and Kate "adored each other." He brought
her to Clontarf for tea and sat adoringly at her feet. Anny was
most hospitable and Ruth very friendly, but Hilde was glum
and withdrawn.

Erwin's siege of Kate lasted almost a year before she capitu-
lated. This event was commemorated in another poem written
early in August, 1945 that began:

> Ich träume von einer Sommernacht
> wo du mir in Armen liegst
> und ich endlich endlich hab vollbracht
> wofür du dich an mich schmiegst.

I dream of a summer night / when you lie in my arms / and at last at
last I have accomplished / that for which you press close to me.

Erwin consoled her with the strange statement: "You had no
father, now you get one to lie in bed with." Whereas Kate was
worried about becoming pregnant, it is possible that Erwin was
consciously motivated by the desire to have a son. The love
affair continued until the pregnancy was evident. Kate confided to Lena Lean that "she did not understand how it ever
happened," a statement that might have been more plausible
in Ireland than elsewhere. The baby was a six and one-half
pound girl, born June 3, 1946, and christened Linda Mary
(pseudonym).

When baby Linda was a few months old, she was taken to
the Schrödinger home, where Anny and Lena devoted themselves to her care. Hilde and Ruth meanwhile had returned to
the March house in Innsbruck. Anny offered to give Erwin a
divorce if he wished to marry Kate and settle down with their
daughter, but neither mother nor father had any interest in
such an arrangement. Erwin was delighted with the child,
however, and loved it dearly. He wrote about the new member

of the family to Hermann Weyl, who replied: "To Anny and Linda a heartfelt greeting and kiss! In this shattered or shattering world the natural love of life and gaiety of such a small creature is indeed like a ray of sunshine and a last hope."

HIROSHIMA

On August 6 came news of the first use of the atom bomb – the extermination of Hiroshima. Schrödinger was dismayed by the slaughter and the new horrors thus introduced into human affairs. In a letter to Weyl, he wrote: "I find the development of things on this planet so desperate that I close my eyes and don't look around . . . The dangerous enemy is the <u>State</u>. The abscess of fascism has been cut out, but the idea lives on in its sworn enemies . . . I shudder at the thought that it can go so far with us, but it has already gone much too far. For example, the atom bomb." Unlike so many other great physicists of this time, Schrödinger never applied his mind to the development of weapons of mass destruction.

He saw at once an interesting statistical problem in the chain reactions responsible for nuclear explosions: what is the critical mass below which the occurrence of spontaneous explosion is effectively impossible? In November, he read a paper to the Academy on "Probability Problems in Nuclear Chemistry." This paper appeared at a time when publication on such a subject was strictly prohibited in most of the world, and it must have aroused considerable apprehension as to how many more nuclear secrets might be derived and independently published by maverick Irish physicists.

UNIFIED FIELD THEORY

Early in 1946, Schrödinger began to attack the unified field theory with fresh enthusiasm. He was encouraged by the renewal of his correspondence with Einstein after a hiatus of more than two years. On January 22, Einstein sent him two unpublished papers: "I am sending them to nobody else, because you are the only person known to me who is not wearing blinders in regard to the fundamental questions in our science. The attempt depends on an idea that at first seems antique and unprofitable, the introduction of a non-symmetrical tensor as the only relevant field quantity . . . Pauli stuck out his tongue at me when I told him about it."

Schrödinger replied in a long letter on February 19. He said that Einstein's work had made a deep impression and he had studied it intensively for three days. Einstein was astonished that he was able to go so thoroughly into "my new hobby-horse" in such a short time. The correspondence was now flying back and forth across the Atlantic. Einstein was seeking singularity-free solutions that would perhaps reveal the origin of particles in the structure of space–time. Schrödinger told him: "You are after big game as I would say in English. You are on a lion hunt, while I am speaking of rabbits." The essence of all that is of permanent value in this correspondence has been distilled by Schrödinger into the final sections of his little book, *Space–Time Structure* (1950).

On May 20, Einstein wrote: "We have squandered a lot of time on this thing, and the result looks like a gift from the devil's grandmother." He wondered if it might be necessary to introduce probabilities into field theory instead of trying to specify the "real situation" of the particles.

Schrödinger replied virtually by return mail: "I have not laughed so much for a long time as over the 'gift of the devil's grandmother.' For in the preceding sentences you described exactly the way of the cross that I also traveled in order to end up with something that is probably even more impossible than your result."

CONTACTS RENEWED

As soon as travel became possible after the war, the Dublin Institute invited scientists from overseas. From March 9 to 21, Pauli came from the Institute for Advanced Study in Princeton to lecture on meson theory, nuclear forces, and elementary particles. His lectures and the resulting discussions brought much new information about unpublished advances in Europe

21. Schrödinger, held in the hand of God, contemplates unified field theory. Painting by John Synge

and America. During May and June, Leon Rosenfeld from Utrecht lectured on nuclear theory. On June 14, Whittaker came from Edinburgh to give a lecture on cosmology and its religious implications. Many of the Dublin dignitaries attended, including de Valera. Afterwards, Msgr. Paddy Browne, in his mischievous way, remarked to Erwin: "I should have liked to have a nap myself, but the Provost and the Taoiseach were both asleep and, after all, the existence of God was at stake."

Anny and Erwin had been looking forward to renewing old friendships in their first post-war visit to England and the Continent. Toward the end of July they crossed the Irish Sea for the first time in over six years, going first to Cambridge, where Dirac had arranged a small meeting of physicists.

Erwin was now anxious to renew his love affair with Hansi Bauer. They had corresponded faithfully, but had much to tell each other that could not be put into words. Leaving Anny in Cambridge, he met Hansi in London. They had only five days together. Hansi now had two children, a boy of eleven and a younger girl, so that she and Erwin could discuss the problems of parenthood. Over Easter, 1947, Hansi came for a visit to Dublin. She met Kate and was not impressed. Apparently she did not meet Sheila, who would have been more of a match for her.

Erwin had lectures scheduled in Switzerland, and on August 1, he picked up Anny and traveled to Zurich. From there he proceeded to Ascona to take part in an *Eranos Tagung*, one of a series of annual conferences founded by Carl Jung, the famous advocate of ancestral memories. The topic of the meeting was "The Spirit and Nature," and Schrödinger's lecture was on "The Spirit of Science." He began with an eloquent statement of the true relation of spirit (*Geist*) to science (*Naturwissenschaft*). Its theme was that "spirit is to an eminent degree subject, and thus evades objective examination." He quoted Sankara's commentary on the Vedanta–sutras: "Subject and object – the 'I' and the 'not-I' – are in their essence opposed to each other like light and darkness." The great Indian philosophers were concerned only with the "Ego that consists of thought" and its relation to the Godhead. Schrödinger wished to identify this "Ego that consists of thought" with "spirit." Science can examine only the object, the non-self, and "the spirit, strictly speaking, can never be the object of scientific inquiry, because objective knowledge of the spirit is a contradiction in terms. Yet, on the other hand, all knowledge relates to the spirit, or, more properly, exists in it, and this is the sole reason for our interest in any field of knowledge whatsoever."

THE EINSTEIN DEBACLE

Soon after his return to Dublin, Schrödinger resumed his letters to Einstein, keeping him informed of progress with the unified field theory by reports at about fortnightly intervals. Einstein's replies were less frequent. On Christmas Day, he wrote to apprise Einstein that he had carried through the variation procedure for a complex field. On New Year's Day, 1947, Einstein replied that "You are a clever rascal [*ein raffinierter Gaunerl*]." Erwin was delighted with this epithet: "No letter of nobility from emperor or king, neither the order of the garter nor the cardinal's red hat could do me greater honor than to be called a clever rascal by you in such circumstances." In fact he was so pleased that he listed all the reasons why Einstein should move to Ireland. "One can live in unbelievable peace and tranquility. This is due to the boundless lack of education and intellectual disinterest of the great majority of the population. That is naturally expressed unkindly. One can also say they are a natural, simple people, who do not go in for humbug." Einstein replied that he could not leave Princeton where they had done so much for him, and besides, a move to Dublin would deprive him of the pleasure of their correspondence. These first days of the new year marked the high point of Schrödinger's love and admiration for Einstein, but their collaboration and friendship were about to experience a disastrous misunderstanding.

Schrödinger was scheduled to read his paper on "The Final Affine Laws" at the meeting of the Royal Irish Academy on January 27. He had been working on this paper for at least six months, but judging from his letters to Einstein, it had not assumed its final form until the second week in January. The day before the meeting, he wrote a long letter to Einstein to outline what he believed was a major breakthrough:

Today I can report on a real advance. Maybe at first you will grumble frightfully, for you have explained just recently that and why you don't approve of my method. But very soon you will agree with me . . . In brief, the situation is this. If in the affine theory, which I have developed in general form in recent years, one takes the special, the only reasonable Lagrange function, namely the square root of the determinant of the Einstein tensor, then one obtains something fabulously good.

An outline of the paper to be presented the next day at the Academy was then given.

Schrödinger was so entranced by his new theory that he threw caution to the winds, abandoned any pretense of critical analysis, and even though the new theory was scarcely hatched, he presented it to the Academy and to the Irish press as an epoch-making advance. "I have the honour of laying before you today the keystone of the Affine Field Theory and thereby the solution of a thirty-year-old problem: the competent generalization of Einstein's great theory of 1915." The complete paper was then submitted to the secretary, and Schrödinger was surrounded by reporters from the Dublin papers who had been alerted to the great scientific event.

The *Irish Press* next morning carried the story:

Twenty persons heard and saw history being made in the world of physics yesterday as they sat in the lecture hall of the Royal Irish Academy, Dublin, and heard Dr. Erwin Schrödinger . . . It was later he told me that "the theory should express everything in Field Physics" . . . The Taoiseach was in the group of professors and students . . . Schrödinger disappeared through the snowy traffic on his veteran bicycle, before he could be questioned further, but later in his Clontarf home, chain smoking, he told me: "It is practically impossible to reduce the theory to terms that the man in the street can understand. It opens up a new field in the realm of Field Physics. It is the type of thing we scientists should be doing instead of creating atomic bombs."

The reporter asked if he was quite confident in his solution, and Schrödinger replied, "This is the generalization. Now the Einstein Theory becomes simply a special case . . . I believe I am right. I shall look an awful fool if I am wrong."

The story of the great discovery reported in Dublin was picked up by the international wire services and flashed around the world. The science editor of the *New York Times*, William L. Laurence, immediately secured photostats of the original paper and of Schrödinger's remarks to the Academy and sent them to Einstein, Oppenheimer, Wigner and others, asking for their comments.

Einstein could hardly believe that such grandiose claims had been made for what was at best a small advance in the work that they had both been pursuing along parallel lines. He devoted great care to preparing a statement in response to Laurence's request. After explaining the preliminary, formal, and purely mathematical state of general field theory, he continued:

Schrödinger's latest effort . . . can be judged only on the basis of its mathematical-formal qualities, but not from the point of view of "truth" (i.e., agreement with the facts of experience). Even from this point of view I can see no special advantages over the theoretical possibilities known before, rather the opposite. As an incidental remark I want to stress the following. It seems undesirable to me to present such preliminary attempts to the public in any form. It is even worse when the impression is created that one is dealing with definite discoveries concerning physical reality. Such communiqués given in sensational terms give the lay public misleading ideas about the character of research. The reader gets the impression that every five minutes there is a revolution in science, somewhat like the *coup d'état* in some of the smaller unstable republics. In reality one has in theoretical science a process of development to which the best brains of successive generations add by untiring labor, and so slowly lead to a deeper conception of the laws of nature. Honest reporting should do justice to this character of scientific work.

Einstein's comment also went out over the wires of the international press, together with a quotation of Schrödinger's remark that "if I am wrong I shall look an awful fool."

On February 3, before he had seen Einstein's remarks,

Schrödinger sent him a long letter in which he offered a tortuous explanation of the overblown newspaper accounts, claiming that he had exaggerated his discovery for the sake of local publicity and an increased budget for the Institute. Such an excuse could hardly have improved his standing with Einstein. It was bad enough to make a mistake and announce a great discovery in an excess of enthusiasm, but to make the claim so as to bring pressure on the government to raise salaries would have been dishonesty.

Actually other evidence indicates that Schrödinger really did believe that he had made a major breakthrough, so that the excuses given to Einstein were concocted later. Thus he had written to Hansi on January 14:

Suddenly it solved itself miraculously, fairylike, in reality it would have been during four or five days . . . barely interrupted by sleep and with considerable consumption of alcohol. After this came frequent thoughts leading the way (most immoral, what?). One would think that a fellow who the night before drank himself to sleep shouldn't be capable of any thinking, and have a hideous hangover. But no – no hangover. Breezily up and a little wash and to the notebook, and it all goes as if oiled.

At my age I had completely abandoned all hope of ever again making a really big important contribution to science. It is a totally unhoped for gift from God. One could become believing or superstitious [*glaubig oder aberglaubig*], e.g., one could think that the Old Gentleman had ordered me specifically to go to Ireland in 1939 as the only place in the <u>world</u> where a person like me would be able to live comfortably and without any direct obligations, free to follow all his fancies.

In any case, the entire episode reveals at best a lapse in judgment, and when he read Einstein's comment, Schrödinger was devastated. Einstein wrote a curt letter on February 2, saying that he believed they had discussed the theory sufficiently, and the time had now come to obtain some rigorous solutions. If anything really was achieved, he would write again to let Schrödinger know.

Schrödinger collected the newspaper files and some of the relevant correspondence into a folder labeled *Die Einstein Schweinerei*, which may be translated as "the Einstein mess."

THE WORLD VIEW OF SCIENCE

After this debacle, Schrödinger began to devote more time to philosophy and less to unified field theory. He had reached his sixtieth birthday, a time to put aside storm and stress and to consider the finals ends of life. He took up Spinoza, Einstein's favorite philosopher, and began to read again the ancient Greeks in a search for the origins of the dichotomy between scientific and religious thought, but he was not inclined to change his view that physics provides no answers to philosophical questions.

During the autumn of 1947, he finished a long essay, "On the Characteristics of the World View of Science," which he sent to *Acta Physica Austraica*, as a sort of intellectual present to a "liberated" Austria. He began by quoting with approval the statement of John Burnet that "science is thinking about the world in the Greek way," and he found that this way was based on two assumptions. (1) Comprehensibility: the belief that natural events can be understood and explained (2) Objectivation: the exclusion of the perceiving subject from the world picture that is to be understood and its relegation to the role of an external observer. But what does it mean to *understand* nature? At this time, almost all professional philosophers of science were devoted to a "received view" called *logical positivism*, which was a restatement of the ideas of Mach in more technical terms. Schrödinger had not spent much time reading contemporary Machians, since he had read every word of their master.

He had never been more than a reluctant disciple of Mach, and in this essay he took a more definite stand against positivism. His first argument was that historical reconstructions provide understanding but they are certainly not merely economical summaries of sense perceptions. Thus positivism cannot provide a general epistemology, although it might still be valid for science. His second argument was based on the structure (*Gestalt*) of fairly complete theories. Many scientific theories, Darwin's theory of evolution for example, cannot be expressed in terms derived from sense perceptions, and thus positivism is not a valid philosophy for all of science. The failure of Mach and his followers to accept the atomic theory in physics and chemistry should in itself raise doubts about the validity of positivism in these fields. The wave particle duality of quantum mechanics is a crucial case; the Machians say that no meaningful "picture" is possible, but this conclusion is more like a defeat than a victory for their philosophy. Within a few years, from about 1950, philosophers of science would begin to abandon the "received view" of logical positivism. Their arguments would be similar to those of Schrödinger, but expressed in more technical language and with more cogent logical analysis.

To emphasize the absence of mind from the world picture, he made a dramatic comparison with Poe's story, "The Mask of the Red Death." When a daring reveler tears the mask and cloak from the dread figure, he finds beneath them – *nothing*. The reason that our perceiving and thinking self is not to be found in the world picture is simply this: *it is the world picture*. It is identical with the whole and thus cannot be found in any part.

How then are we to consider the *apparent* multiplicity of selves? One answer would be a multiplicity of worlds, the "horrible doctrine of Monads due to Leibniz, each a world to itself, without windows, in agreement through a pre-established harmony." The opposite alternative is the unification of consciousness. Schrödinger concludes that the mystical experience of union with God leads regularly to this understanding, and he quotes the words of the Persian poet Aziz Nasafi: "The world of spirits is a single spirit standing like a light in back of the world of bodies and shining through each individual that comes into existence as through a window. According to the kind and size of the window, more or less light penetrates into the world. But the light always remains the same."

From his time as a young man in cold and hungry wartime Vienna, when he delved deep into the Upanishads, through his years of great scientific accomplishment, to his situation as a philosopher on the verge of old age, Schrödinger had never deviated from a religious understanding of our mysterious world. His position was captured in a remarkable painting by his Dublin colleague John Synge, who also claimed to be an atheist, which shows Erwin held in the hand of God as he ponders the equations of unified field theory. Yet Erwin could never be a believer in any dogmatic religion – his search for truth could never reach a conclusion. As Hansi once said, "With him nothing was ever fulfilled. He was always hoping for the ultimate. High expectations by great men cannot be fulfilled."

A YEAR OF CHANGES

As Erwin reviewed the year 1947, he found it to have been one of the happiest in his life, in large measure owing to his love for his baby daughter and her response to him. The new year was to bring both sorrows and joys. It began with some improvement in the financial situation of the Institute. The annual budget was increased to £9,370, which included three senior professors at £1,500 each.

Times were hard in the first years after the war, and there was a widespread feeling that a change from the Fianna Fail government was overdue. As part of the political skirmishing, the Institute came in for its share of attacks. For example, the *Irish Independent* commented:

There are some people in this country whose minds can never rise from the level of practical finance into the realms of theoretical physics. For all we know there may be some who are so lowbrow and so selfishly concerned with the vulgar problems of dry bread and wet turf that they do not even know what theoretical physics and cosmic physics are . . . Of one thing we feel sure – that Irish is not a compulsory subject in these schools, for any such rule would be a discourtesy to the miniature League of Nations which labours for dark Rosaleen – at dark Rosaleen's expense of course – in this modest seat of learning.

De Valera called a general election, and the result was that his party was outnumbered in the Dail by a coalition of smaller parties, and on February 18, by a vote of 75 to 70, John

Costello was elected Taoiseach. For the first time in sixteen years, Dev took a place on the back benches as leader of the opposition.

On February 17, Erwin and Anny became citizens of Ireland, in a simple ceremony, which consisted of swearing an oath of loyalty to the Republic before a judge at the Four Courts. It was a fitting gesture of appreciation to the country that had offered him and his family a hospitable refuge during the worst of times. His reasons, however, were more practical, being based on the difficulties of securing travel documents from the Austrian government while it was under the control of the four allied powers. He did not wish to give up Austrian citizenship, for he thought his rights to a pension might depend upon it, but he found that it would be possible to have dual citizenship, so that in becoming Irish, he would remain no less Austrian.

During February, Schrödinger delivered four public lectures at University College, Dublin, on "Nature and the Greeks," and they were repeated May 24–31 as the Shearman Lectures at UC London. *Nature and the Greeks* is one of his finest works, probably better in the written version than as heard in lectures, since the density of ideas and the elegance of their expression deserve several readings.

His private views of organized religion had not changed appreciably. Thus he wrote to Benjamin Farrington, the historian of Greek science, in an analysis of the ideas of Lucretius:

To make the extirpation of religion one's first and principal aim is not very wise anyhow. I believe that a man like Lucretius realizes only half the strength of the enemy he is up against. The hypocrisy of the governing classes who promote superstition as an efficient means of ruling over the dispossessed has a strong ally in the religious desire of many men and women, especially among the poor and unhappy. Lucretius does not seem to have much understanding of this side of human nature. He knows about serious religious longing as little as he knows about the shape of our planet.

After the London lectures, Erwin returned immediately to Dublin, for Anny was suffering from a severe depression. She had been grieving over the departure of Ruth, whom she dearly loved. After she made a fairly serious attempt at suicide by slashing her wrists, she was admitted to St. Patrick's Hospital on June 17. Her psychiatrist was Maurice Drury, the great friend of Wittgenstein. He prescribed a short course of electroshock treatments, after which she improved considerably and was able to leave the hospital in a few weeks. For the next eight years, she suffered from recurrent attacks of depression, and when they became too bad, she would get into her car, drive to the hospital, and sign herself in for treatment. She never again attempted suicide and bore her illness with understatement and stoicism.

Erwin also had a medical problem. Cataracts in both eyes had been growing more troublesome and now he had difficulty in reading even with strong glasses. He consulted Louis Werner, Dublin's most distinguished ophthalmologist, who scheduled the cataract extraction on the right eye for June 29. Dr. Werner recalls that "Erwin approached his operation not with fear but almost with pleasurable anticipation as if we were jointly carrying out an experiment in optics . . . He was a well behaved patient although I suspected that occasionally he disobeyed orders and lifted up his bandage to have a peep around." The cataract in the left eye was extracted by Dr. Werner a year later.

The operations were successful and Erwin's vision was restored to near normal with corrective lenses. He wrote to thank Dr. Werner: "The world is full of beauty again, which I owe to your masterful knowledge and skill . . . You have made me a new man who enjoys life once again . . . More than ever before I realize now what a human being actually is: a pair of eyes – with something around them to keep them going, and also to turn their giving into 'mind-stuff,' in an entirely miraculous and unfathomable manner. All the rest of our sensing, however relevant, is yet only ancillary to sight." Dr. Werner refused to accept any fees, and Erwin later dedicated to him his book *Space–Time Structure*.

With Anny not well, the care of baby Linda became primarily the responsibility of Lena Lean, but they also had a maid, Molly, to help. Kate visited frequently, but her relations with Erwin were less than cordial. She often threatened to take the baby away, and might have done so if her mother had not been so fearful of ever being seen with it. The Schrödingers had decided to legally adopt the baby.

In the late summer of 1948, while Erwin and Anny were away in England, Kate met Lena one day walking the baby in its pram; she took it away and there was nothing Lena could do to stop her, since, after all, she was the mother. Erwin was in England and the first news came in a telegram brought by an excited messenger, "Baby abducted from pram. What shall I do?" Kate took Baby Linda about as far as she could go from Ireland, and Erwin never saw his daughter again, although he contributed to her support. As a girl she never heard her father mentioned until many years later, when Ruth established contact with her half-sister. Kate never married, perhaps in obedience to a vow she had made to herself before becoming Erwin's lover.

Early in August, Erwin and Hansi set out on a short tour of North Wales. They stayed for a while at the famous seacoast hotel at Portmeirion, renting a chalet called "The Golden Dragon." Bertrand Russell was at the hotel with his eleven-year-old son Conrad. One afternoon there was a garden party at the house of the Welsh philosopher Rupert Crawshay-Williams. They started drinking tea from mugs and then filled them with whisky to encourage the flow of philosophical ideas. Schrödinger expounded the absolute identity and hence non-existence of elementary particles, but most of the time they all listened to Russell who was in sparkling form at the age of seventy-six.

Meanwhile, Erwin had found a new friend in England, Lucie Rie. Lucie was born in Vienna, the daughter of a Jewish doctor. She attended the Vienna Gymnasium and the School of Applied Arts, where she studied with the famous potter Michael Powolny. In 1938, at the age of thirty-six, already with a considerable reputation in her field, she escaped from Austria to London. She became one of the most influential studio potters of her generation, providing British artists with a modern alternative to the folk art and imitative orientalism which had been the prevailing fashions.

Hansi introduced Erwin to Lucie while he was in London for

the Shearman lectures and they were immediately fascinated by each other – he saw her almost every day during that visit. When he returned to Dublin, Lucie wrote inviting him to stay with her during his next visit in August. Hansi was not much inclined to jealousy, but as Erwin began to spend more time with Lucie, his relations with Hansi naturally became less intimate. For Erwin, the visual sense was all important, and both these women were professional artists, who had much to teach him. With them he experienced a fusion of romantic and intellectual attractions.

PHILOSOPHICAL YEARS

Schrödinger attended the Eighth Solvay Congress in Brussels from September 27 to October 2. As in the first conferences after the First World War, no scientists from German universities were invited, but many old friends were there, including Bohr, Pauli, Dirac, Kramers, and Lise Meitner. The principal topic was "Elementary Particles." At this meeting Schrödinger must have thought that the work he had been doing for the past ten years was far from the concerns of all the other theoretical physicists.

After he returned to Dublin, his research work moved even more definitely from mathematical physics to philosophical problems concerning the nature of the physical world and our knowledge of it. There were a number of new fellows in the School of Theoretical Physics, which was indeed like a league of nations. Walter Thirring was already one of the most promising of the new generation of theoreticians and would eventually succeed his father in the Vienna professorship; his brother Harald had been killed on the Russian front. Ernan McMullin, recently ordained, had studied physics at Maynooth.

Erwin was still an enthusiastic hiker. Almost every Saturday a group of ten or twelve would meet at the Institute, pile into cars, and drive about 20 miles into the Wicklow Mountains, where they would set forth on a strenuous walk. They would usually end up late in the afternoon at some small country pub for supper washed down with Guinness or ale. After supper, the talk almost always got around to religion. Erwin would present "his variety of pantheism," Eva Wills would take the Presbyterian side, and McMullin was expected to defend Catholic orthodoxy. The talk went on till long after dark, when they would finally drive back down the mountains. McMullin recalls that at this time Schrödinger was "far too interested in religion to be called an atheist" – even when he was consulted about physics, the conversation eventually came around to religion.

On May 12, 1949, Schrödinger was elected a foreign member of the Royal Society of London. His election had been unduly delayed by complex politics. No German or Austrian citizens were elected from 1938 to 1948, but his actual citizenship was not the problem. Until 1948, the regulations were that foreign nationals living outside His Majesty's Dominions were eligible for foreign membership, but from 1948, all foreign nationals were eligible. To have elected Schrödinger before 1948, would have been tacitly to admit that Ireland was not a British dominion. The Costello government passed the Republic of Ireland Act in 1948, severing the last links with the British Commonwealth, but not in fact with the Royal Society, since citizens of the Irish Republic are still eligible for election to fellowship.

GEDICHTE

In 1949, for the first time since 1923, Schrödinger published no scientific articles, but a small book of his poems appeared under the imprint of Helmut Kupper in Godesberg. Most lovers of German poetry have deprecated these poems. As might be expected from someone who had no music in himself, there is often a lack of euphony and a crabbed feeling in the language. Technically they are proficient, especially his mastery of the Petrarchan sonnet. Aside from the love poems, which became almost a standard accessory in his amorous adventures, the poems express two principal themes, his love of nature and a metaphysical desperation often bordering on despair. There is a great sincerity in the poems of this latter kind, in which he abandons the romantic conventions and struggles with the meaning of his existence. Thus one can forgive his imperfections as a poet for the sake of the honesty of his self-revelations.

As an epigraph for his *Poems*, Erwin chose a quotation from Goethe's *Egmont*, "*Himmelhoch jauchzend, zum Tode betrübt* (Exultant unto heaven, dejected unto death.") It is from the song of the heroine Klärchen, which continues "*Glücklich allein, ist die Seele die liebt* (Alone is happy, the soul that loves)." Besides expressing the oscillation of the poems between extremes of joy and sadness, the quotation is interesting in that it calls attention to the character Klärchen, who gains the love of the noble prince Egmont despite her inferior social standing and the opposition of her determined mother, and who chooses to die rather than live without him.

From childhood, when he presented his mother with a little book of his poems, to old age, when he set forth his world view of Vedanta in poetic form, Erwin was often inspired to reveal his inmost doubts and longings in his poetry. As he told Max Born, appreciation of one of his poems gave him much more pleasure than any praise of his scientific papers.

NEVER AT REST

The biography of Isaac Newton by R. S. Westfall was called *Never at Rest*, which suggests the ceaseless activity of the mind of the great scientist in its perpetual quest for deeper understanding. A similar restless striving motivated Erwin Schrödinger in the later years of his life. In lectures, essays, radio talks, and perhaps most of all in his letters to other scientists, he would select various fundamental ideas and concepts and try to reach some final conclusions about them, or at least to find answers that satisfied both his reason and his intuition.

He wrote a long letter to Sommerfeld for the eightieth birthday of this revered old friend.

I do not believe that we can approach any understanding of the "mind–matter" problem on a dualistic basis. There is no reason at all for dualism. Matter is a construction from sense impressions and representations in a certain combination, and what one calls "an individual mind" consists of course of the same elements. It is the same material, merely comprehended in a different way. In the first kind of comprehension the world is finally constructed, in the second, the self. The world is certainly no dream, no phantasm, and certainly the self is not merely a "summation of connected sensations in a peculiarly fluctuating sequence" . . . When I reflect carefully, then two things are not given to me at all. Both come to me uniformly from the same source, I find no inhomogeneity in the flow, no separation of spiritual and material – it is all from the same substance.

The dualists could contend that "knowing is mind and being is matter" and the interface between the two is "consciousness," but Schrödinger had by now definitely discarded any such view.

Schrödinger's most important work in 1950 was his 119-page book *Space–Time–Structure*. Into these pages he compressed all that he had learned about the geometry of space–time and its affine and metric connections. The book has been the *vade-mecum* of two generations of students taking up the subject of general relativity. It will remain one of the classics of science even though Schrödinger and Einstein, as viewed from our present-day perspective, ultimately failed in their efforts to derive electromagnetism from the structure of space–time.

A CLOSE CALL

After Heitler resigned, to accept a chair at Zurich, Schrödinger became Acting Director of the School and actually showed considerable administrative ability in drawing up procedures for its more efficient functioning. Cornelius Lanczos, a Hungarian Jew and a great expert in relativity theory, came in 1952 as visiting professor and stayed as a senior professor and later director. He was a small man with an aureole of white hair which made him look like a retired angel. Like Erwin he was interested in young ladies and in the theater, and he would often invite the waitress who had served his dinner to accompany him to the play. He sat in the front row at lectures and if he fell asleep, Erwin would stealthily approach and suddenly raise his voice, causing Cornelius to awake with a start. Erwin was quite fond of him but this feeling was not reciprocated.

In mid-September 1952, Schrödinger wrote to John Synge from the Tirol. "I am feeling quite . . . *klein und hässlich* meaning guilty and humble for leaving you alone for so long. Anyway I have been having and am still having a glorious time, and I ought to return filled with so much energy and vigour that it may easily blow up the roof of Old Merrion Square the moment I enter the students' room in the top floor."

The British Society for the History and Philosophy of Science was planning a meeting in London in early December to discuss the interpretation of quantum mechanics. They had invited Bohr, Popper, Born, Schrödinger and others, and everyone was looking forward to exciting confrontations between the upholders of the Copenhagen orthodoxy and their critics. Max Born wrote to Erwin that it would be like the famous 1895 Lübeck conference on atomism, and he did not know whether he would be cast as the matador or the bull, but still he was confident that their friendship would survive the battle.

Schrödinger returned to Dublin the first week in October, feeling in better health and more full of energy than he had been for some years. Three weeks later he was stricken with appendicitis, which he tried to ignore during the first few days. When he was finally rushed to a hospital, the inflamed appendix had burst and his condition was critical. The surgeon performed an emergency operation. Fortunately, antibiotics were by now available and the infection was contained. Erwin began a gradual recovery, but it would be three months before he was fit for serious work.

At the age of sixty-five, however, his body lacked resilience, and never again would he feel the almost youthful energy and high spirits of that summer in Austria. The annual attacks of bronchitis now became more wearing. Even when confined to bed with a racking cough, he would fill the room with smoke from his ever-present pipe.

ARE THERE QUANTUM JUMPS?

The paper that Schrödinger would have given at the London meeting was published later in the *British Journal for Philosophy of Science*. It is a rather rambling account of his ideas on a variety of topics rather than a detailed analysis of the occurrence of discontinuities in physics.

The problem of quantum jumps can be seen most clearly in the theory of spectroscopic transitions. Suppose that a hydrogen atom absorbs a quantum of energy and undergoes a transition between two stationary states with eigenfunctions Ψ_1 and Ψ_2 having energy eigenvalues E_1 and E_2. Schrödinger attacked the idea that such a transition can occur *instantaneously*. (One current view was that the transition is indeed instantaneous but the time at which it occurs cannot be specified exactly owing to the uncertainty principle in the form $\Delta E \Delta t = h/4\pi$.) He was correct in this criticism and, as Heisenberg pointed out, the atom passes through a series of superposition states, $\Psi = c_1 \Psi_1 + c_2 \Psi_2$, where c_1 and c_2 are continuous functions of t. These superposition states are not eigenfunctions of the energy operator H, but may in certain cases be detectable as eigenfunctions of some other observable. According to Heisenberg, it is the act of observation that precipitates a jump into either state 1 or state 2.

As Heisenberg commented,

Schrödinger therefore rightly emphasizes that . . . such processes can be conceived of as being more continuous than in the usual picture, but such an interpretation cannot remove the element of discontinuity that is found everywhere in atomic physics: any scintillation screen or Geiger counter demonstrates this element at once. In the usual interpretation of quantum mechanics it is contained in the transition from the possible to the actual. Schrödinger himself makes no counterproposal as to how he intends to introduce this element of discontinuity, everywhere observable, in a different manner from the usual interpretation.

It should be added that quantum mechanics itself, whatever its interpretation, does not account for the transition from the "possible to the actual."

THE MASS OF THE PHOTON

Schrödinger's aversion to discontinuity may be seen also in the last substantial work that he did in Dublin, in collaboration with Ludvik Bass, a post-doctoral scholar from Prague and Vienna. In "Must the Photon Mass be Zero?" they wrote "In a reasonable theory we cannot admit even hypothetically that a certain type of modification of Maxwell's equations, however small, would produce . . . grossly discontinuous changes."

Maxwell's equations are singular in admitting only transverse electromagnetic waves *in vacuo*. In the nineteenth century this was interpreted as incompressibility of the luminiferous aether, in the twentieth century as the vanishing of the photon rest-mass. An arbitrarily small but finite rest-mass m would permit the existence of longitudinal waves and photons, and hence it would introduce a third independent direction of polarization. In thermal equilibrium, as in black-body radiation, the three directions of polarization would have equal shares of momentum and energy by equipartition over the degrees of freedom. Radiation pressure and the constants in front of Stefan's and Planck's laws should then have values $3/2$ times those actually found. Does this prove that the photon mass must in fact be precisely zero?

Bass and Schrödinger considered black-body radiation in a cavity made of a perfect conductor. They found that even at the upper limit of $m = 10^{-41}$ g, a cavity with walls as thick as the earth could not confine longitudinal photons, and an appreciable fractional conversion of transverse to longitudinal photons would take longer than the age of the universe. Thus any such longitudinal photon must lead a ghostly existence. The conceptual jump in the properties of black-body radiation does not occur because the relaxation time to equipartition over the three directions of polarization goes to infinity as the photon mass goes to zero. Thus the photon mass is not *required* to be absolutely zero, and the question posed by Bass and Schrödinger continues to attract theoretical interest.

CHAPTER 12

Home to Vienna

Soon after the end of the Second World War, Schrödinger's friends in Austria began to explore various avenues for his return to his native land, but it was to prove a long and tedious process. As early as February 1946, he agreed to accept a professorship at Vienna, provided he could find a suitable successor at DIAS. In April, the Austrian authorities advised that he should apply for a restoration of his professorship at Graz, and then they would arrange a transfer to Vienna, but he was not willing to do this without a more definite understanding.

The main problem was the continuing occupation of Austria by the four allied powers. The winter of 1947 was unusually harsh and there were food riots in Vienna recalling those of 1919. In May an attempted communist coup failed and in June, Austria was included in the Marshall plan, receiving more dollars per capita during its first year than any other country. In February 1948, communists took control of Czechoslovakia, and Stalin showed no willingness to conclude an Austrian peace treaty. Thus, although Schrödinger wished to return to Austria, he made it clear that he had no intention of doing so while the threat of a Russian takeover remained. As the peace negotiations dragged on, they became increasingly acrimonious.

The situation at the Vienna Physics Institute was rather surprising. As the Russians approached in 1945, the Nazi boss of the Institute, Georg Stetter, fled to Egypt with his assistant Orthner, and Hans Thirring was recalled to his old position. By now he was more concerned with the peace movement than

with physics. Przibram who had escaped to Belgium at the
Anschluss returned to his old chair but soon retired. After a few
years, Thirring brought Stetter back, and soon Orthner was
also recalled. Thirring extended his pacifist philosophy even to
the former Nazis, and he believed that old enmities should be
buried and a fresh start made, a feeling widely shared by
Austrian politicians and intellectuals. Nevertheless, the old
political divisions between the socialists and the clericals were
soon revived, but with 90 percent of Austrian Jews either exiled
or exterminated, the blacks were bereft of their favorite
slogans. Both parties were striving to attract the former Nazis
who had comprised the great majority of the population.

SABBATICAL IN INNSBRUCK

The University of Innsbruck, at the suggestion of Arthur
March, invited Schrödinger to spend the winter term, through
March 1951, as a visiting professor, and he gladly accepted.
Innsbruck was in the French occupation zone and so presented
no political hazards. He was happy to be again in the place he
loved best in the world, the mountains of Tirol, and with
March's assistant Ferdinand Cap and other young colleagues
he made frequent excursions to favorite spots in the environs of
Innsbruck. Not the least of his pleasures was reunion with
Hilde and with his daughter Ruth, now a young lady of
sixteen.

While in Innsbruck, he was asked by the University whether
he would favorably consider a permanent appointment there,
and his response was positive. After considerable correspondence with the University and with the Education Ministry in
Vienna, in which agreement appeared to have been reached,
he was surprised to receive in April 1953, a letter from the
Cabinet Director, which stated that no appointment in Innsbruck would be possible because that university had no open-
ings in physics. He replied that this was for him "a hard
disillusionment whereby I must bury the hope I had cherished
for many a year, that the mountain air of Tirol would provide
a stimulus for the last decades of my work, which declines with

advancing age." He expressed his astonishment that the grounds cited for the refusal must already have existed at the time he was given the invitation. This shabby treatment of Austria's most eminent living scientist makes one suspect that some influence of the old Nazi regime may have lingered on in Vienna.

The Viennese bureaucracy did ask if he would accept a professorship there. Schrödinger's answer was: not under the present political conditions. He had good reason for his fears. Vienna was surrounded by the Russian zone and although it was easy to enter, it was not so easy to leave. On one occasion Erwin and Anny were hauled off the bus at the Russian check-point at Semmering, and only the immediate protest of Hans Thirring to the Russian commandant in Vienna prevented an unpleasant incident.

ALPBACH

After his sabbatical at Innsbruck, Schrödinger spent as much time in Austria as possible. He became an active participant in the summer conferences at Alpbach, which brought together intellectuals for discussions of interdisciplinary topics. Alpbach is a beautiful Tirolean village at an altitude of about 1,000 meters, situated in a sunny valley between the mighty Zillertal and Kitzbühel Alps. It became for Erwin his favorite place on earth, and it was increasingly difficult for him to return to rainy Dublin at the end of his holidays. He wrote that "sunshine is a lovely thing. You cannot really feel that you have enough of it before you feel that you have had too much of it."

Lise Meitner came for a visit and they took her for a five-day tour of the neighboring mountains. At seventy-six, she was tireless; Anny would drop back with her asthma after about 150 meters, while Erwin would plod slowly behind them both, hardly able to get up a small hill. When they returned to Alpbach, he became seriously ill, but refused to see a doctor. As Anny reported, "he imbibed more freely of the good Austrian wine than was good for him and in between he drank ice cold beer and you know he never said no." In early September she finally persuaded him to see Dr. Ursinn, who diagnosed emphysema, bronchitis, high blood-pressure, and a weakened heart. He was advised to get back home as quickly as possible and to go to bed with absolute rest for at least ten days, with no smoking and no alcohol.

After the prescribed rest, Schrödinger improved remarkably, and he was able to go to the Institute for a short time each day. He drank only a mixture of milk and soda water and went to bed at 9:00 p.m., but he was unable to break his addiction to smoking.

FAREWELL TO DUBLIN

Despite his miserable health, Schrödinger continued to work with a concentration and efficacy that would have done credit to a man in his prime. From 1954 to 1956 he published fourteen papers, and he completed another of his elegant short monographs, *Expanding Universes*. He was delighted with a new group of young scholars at the Institute, especially with Bruno Bertotti from Pavia, with whom he formed an enduring friendship, for the lively young Italian shared his interests not only in relativity theory but also in philosophy and poetry.

Anny's visits to Lucan for therapy had now become routine. Her attacks of depression always seemed to begin early in December, aggravated perhaps by the Advent season and mourning for a distant Catholic childhood when the message of salvation was more easily decipherable. Her stays at the sanitarium, however, never coincided with the times when Erwin was seriously ill and required her attention. Typically she would return from Lucan just in time to put him to bed. Once Christmas was over, they both tended to revive and to look forward to springtime and then, in early summer, a return to the Tirol.

In April 1955, Schrödinger became quite depressed, for he was now convinced that his unified field theory was no longer tenable. His correspondence with Einstein had resumed in 1950, but both theorists had become discouraged.

He was pleased, however, with plans for his return to

Austria. After the death of Stalin in March 1953, Khrushchev had decided to resolve the Austrian situation, and in February 1955, he instructed Molotov to come to an agreement that would guarantee the neutrality of Austria after the Swiss model. He wrote to Raab, the Austrian Chancellor: "Follow my example and turn communist, but if I really can't convince you, for God's sake stay as you are." The final treaty was more favorable to the Austrians than versions rejected by the Russians, and the war-guilt clause was removed. On May 15, 1955, the treaty was ceremoniously signed in the Belvedere, and on November 5, the last foreign soldier left Austria. After almost ten years, the way was at last clear for Schrödinger to return to a professorship in his native land.

In early June, Erwin and Anny traveled to Italy, where he gave a paper on the photon mass at a meeting of the Italian Physical Society in Pisa. They remained a week in Tuscany, visiting Lucca, Sienna, Orvieto, and Firenze. In mid-July Schrödinger received official notification of his appointment in Vienna. They created for him an *Ordinarius Extra-Status* from January 1, 1956, with full pension rights upon retirement. He accepted immediately with great joy.

Their last winter in Dublin was difficult, although Ruth came for four months to help with the move to Vienna. Anny was in and out of Lucan and Erwin suffered from phlebitis and then an unusually severe bronchitis. He became so depressed that on Sunday night, February 11, he took four different kinds of sleeping pills and washed them down with whisky. On Monday morning, Anny had difficulty in rousing him and summoned Dr. Dempsey, who said "I've never seen him like that. He nearly killed himself." Dempsey wanted to call in one of the psychiatrists from Lucan for a consultation but Erwin vehemently refused. Under Anny's care he made a good recovery.

HOME AGAIN

By the time they boarded the cross-channel steamer, Erwin was exhausted by the farewell parties in Dublin, lunch with de Valera, lunch with the President, and a large party with colleagues from the Institute. They finally set out on March 23, and Dev came down to the pier to see them off.

They reached Vienna at 4:00 p.m. on April 1 and at 5:00 p.m. the avalanche of welcomes began. They were staying at the Pension Atlanta very near the Physics Institute. Their rooms were filled with flowers, including a large purple azalea with a card reading "The Academic Senate warmly sends greetings to the great son of Austria upon his return home." There was a mountain of letters and telegrams of welcome and congratulation. They were besieged by reporters and photographers, and the stories and interviews with Austria's greatest scientist were on the front pages of all the newspapers.

At noon on April 13, Schrödinger delivered his inaugural address in the Auditorium Maximum of the University. The platform was adorned by civic and academic dignitaries and the great hall was overflowing. Among the guests were Dr. Mayer, director of his old school, the Akademisches Gymnasium, Erwin's cousin Dora Halpern, with whom he used to play and sometimes fight as a little boy, and his early great love, Felicie Bianchi. After a few words of greeting from Dean Laska of the philosophical faculty, Erwin approached the lectern, and as he picked up the spray of narcissus that was lying there, he was greeted by thunderous applause, which continued for several minutes. It was an historic occasion and his pleasure and emotion were evident, but they did not detract from his urbane style as he delivered one of his most polished lectures on "The Crisis of the Atomic Concept." He said that the materialistic picture of the world has been severely shaken and is today more uncertain than ever before, while modern physics is undergoing a revolution, the duration of which cannot be foreseen. Experiments have shown that waves are real and even fundamental entities, which can be directly and even easily observed, which can be said for only a few things in our world. Many physicists today are prepared to consider matter as simultaneously a wave and a particle phenomenon, but one must be wary of a physics that is too philosophical, since it may lead to specious solutions and

22. Inaugural lecture in Vienna, April 1956

prevent a deeper understanding. At the conclusion of his inaugural address, he was given a standing ovation.

On May 5, Schrödinger was awarded the Prize of the City of Vienna at a ceremony in the Festival Room of the Rathaus, which was followed by an elaborate lunch hosted by the Bürgermeister. Erwin enjoyed such occasions although he was supposed to be on a strict diet. Other festivities occurred in June when Lise Meitner came to Vienna for a celebration of the fiftieth anniversary of her doctorate from the University.

It was not easy to find a suitable dwelling in Vienna but the Schrödingers finally located a flat at Pasteurgasse 4, about a kilometer from the Physics Institute. There were five rooms on the third floor, and the building had a lift. Vienna had never seemed so beautiful as during those first months after their return home; the Burgtheater, the Opera, the Ringstrasse, Kärntnerstrasse, Graben, the splendid monuments, and the elegant shops, all bathed in the warmth and sunshine of early summer. They often went to the theater; once the Minister of

Education sent his tickets in the first row of the parterre circle for a play by the old favorite Grillparzer, and as they drove there in their pretty car, Erwin exclaimed "*Na – so fein war ich in meinem ganzen Leben noch nie!*"

But intimations of mortality were never far away. As they were walking in the courtyard of the University, looking at sculptures of its famous professors from bygone times, Erwin suddenly said, "Let's get away from here – I feel as if I were walking in my future cemetery."

He lectured twice a week on General Relativity and Expanding Universes, subjects that had received little attention in Vienna for many years. His teaching was more a discussion than a formal lecture course; he invited the rather small number of auditors to interrupt him at any point that was not perfectly clear, and he would then elucidate with careful detail. There was also a weekly seminar but he thought that "this was more like a higher kindergarten, not like the Dublin Seminar."

MIND AND MATTER

As soon as summer term at the University ended, Erwin and Anny moved to Alpbach, and after a few weeks of mountain air, they were both feeling remarkably well. In early September, however, there was a recurrence of his heart weakness, and he was forced to cancel a planned visit to Cambridge in October to deliver the Tarner lectures. Fortunately, John Wisdom, the professor of philosophy, agreed to read the lectures from Schrödinger's manuscript. The book *Mind and Matter* was published in 1958. It is a fitting sequel to *What is Life?* but it has not had such a far-reaching influence, perhaps because psychobiology is still at such an early stage of development.

The first lecture was on "The Physical Basis of Consciousness." "The world is a construct of our sensations, perceptions, memories. It is convenient to regard it as existing objectively on its own. But it certainly does not become manifest by its mere existence." He is thus led to ask whether one can believe that the existence of the brains of higher animals is a "necessary condition for the world to flash up to itself in the light of consciousness. Would it otherwise have remained a play before empty benches, not existing for anybody, thus quite properly speaking not existing? This would seem to me the bankruptcy of a world picture."

Schrödinger proposes that consciousness is associated with *learning* by a living substance. Living systems, whether they are animals or plants or bacteria, are associated with consciousness whenever they represent the emergence of something new. He suggests further that such consciousness is intimately associated with organic evolution.

A lecture on "The Principle of Objectivation," restated a basic principle that has often been misunderstood by biologists and psychologists. The false conclusion that I am part of a real world that itself arises from my consciousness lets loose a "pandemonium of disastrous logical consequences." One of these is the "fruitless quest for the place where mind acts on matter or vice-versa." The material world can be constructed only at the price of excluding the self, the mind, from it. Mind has constructed the outside world of the natural philosopher out of its own substance, but it could not cope with this awesome task except by paying the high price of withdrawing itself from its own creation. Some interpretations of quantum mechanics have suggested that the distinction between subject and object has broken down even as a basis for physics. He refuses to say that the barrier between subject and object has "broken down" since this barrier does not exist.

The reason why our feeling, perceiving, thinking Self occurs nowhere in our scientific world picture can easily be expressed in six words: It is itself that world picture. The "arithmetical paradox" is that out of many conscious Selves only one world is concocted. Schrödinger sees only one solution: the unity of all Selves in one consciousness. Reality is not to be found in a material world that is outside consciousness; it is consciousness itself. This is the same view of the world that he wrote down in Part One of *Meine Weltansicht* in 1925, at this time not yet published, and it does not differ appreciably from the conclusions he reached as a young researcher in Vienna in 1920. From youth to old age, he has made no essential change in his philosophy. In this lecture he paid tribute by frequent quotations to the work of Charles Sherrington, *Man on His Nature*, and thus may be said to have joined the society of latter-day gnostics for whom this book has become a basic text.

Finally, he considered questions concerning "the other world" and "life after death." He disavowed any attempt to try to answer such questions, and wished only to examine what might be relevant concerning them in science and philosophy. There are truths that do not depend on space and time, for example, certain mathematical theorems. Kant showed the ideality of space and time; they are categories imposed by the mind. But Kant also taught that this *thing*, mind or world, could in principle be capable of other forms that do not depend on space or time. This declaration of freedom leaves open the way to religious beliefs, while preserving them from contradictions arising from either science or philosophy. A mode of existence without time would of course render the concept of "after" meaningless, and "life after death" would not be a meaningful expression. The work of Einstein has even set limits to the concepts of "before" and "after" in the physical world, and this mutability of time is consistent with the idea that the time scale of human life is of little importance.

Schrödinger thought that the statistical interpretation of time, as elaborated by Gibbs and Boltzmann, is even more significant than relativity theory in this context, for it shows that the direction of time is dependent upon a sequence of events that we observe. What we construct in our mind cannot have a dictatorial power over our mind itself, neither the power to call us into life nor the power to destroy us. Thus the statistical theory "frees us from the tyranny of Father Kronos," and we see that Mind cannot be destroyed by Time.

A HARD WINTER

Old age is itself a burden, old age and ill health are worse, and old age, ill health, and poverty are the worst of all. The

Schrödingers were spared only the last of these afflictions. As he entered his seventieth year, Erwin wrote that "the only thing I have enough of now is money." When they returned to Vienna for the beginning of the winter term 1956–57, he was in no condition to withstand the cold, wet weather with his already weakened heart and lungs. In retrospect, Anny at least realized that they should have fled to a warm climate, but there was only one more year to go before retirement and Erwin resolved to see it through. There is little doubt that he would have received an adequate pension in any case, but it was not in his nature to surrender when faced with a challenge to his willpower.

They had excellent medical attention, the famous psychiatrist Hans Hoff, the heart specialist Professor Kurt Polzer, and a physician in general practice, Joseph Schneeweiss, who was always prepared to make house calls. Anny's worst fear was that the arteriosclerosis would affect Erwin's brain. She would drive him to the Physics Institute for his lectures and collect him when he had finished; after a lecture he was sometimes drowsy and incoherent, but the evidence of his letters and other writings at this time shows that his intellectual clarity was not affected by these episodes. Polzer was able to effect a marked improvement in the cerebral symptoms although the underlying disease continued its relentless course, and despite all efforts, Erwin could give only about half his scheduled classes.

Every Saturday morning Polzer and Schneeweiss came, and after an examination and discussion. Erwin would protest, "Herr Colleague, I cannot accept the fact that you devote so much of your valuable time to me," and Polzer would reply, "Herr Professor, you must allow this, that we do everything for you that we have done for our dear Bundespresident and will do much more." This was not overly encouraging, since President Korner who had suffered a stroke in July died on January 4.

Although Schrödinger was officially treated well at the University, he was not so pleased with the physics department. His friend Hans Thirring retired, and there ensued the usual political maneuvering about his replacement. The various factions were agreed only on one thing: to prevent Schrödinger from using his enormous prestige to influence the decision. This academic politics did not bother him so much, however, as did the failure of the department to provide adequate office facilities.

There were compensations in his family life. In May, Ruth had married Arnulf Braunizer, a member of a strongly anti-Nazi family, which for a long time had hoped for a restoration of the monarchy. Their first child was expected in February.

On February 19, Ruth came to Vienna to await the arrival of her baby. Arnulf was anxious to get her away from the March house, where Arthur was mortally ill and Hilde was in a state of nervous exhaustion. Anny was happily busy getting things ready and Erwin was looking forward to his first grandchild. On February 27, Anny and Erwin drove Ruth to the University Clinic, and then spent a sleepless night making periodic telephone inquiries. Labor began at 6:00 p.m. and at 12:35 p.m. a strong boy came into the world. He was called Andreas. Arnulf arrived the next day. Erwin was delighted with his grandson but unfortunately was not well enough to enjoy him as much as he had hoped.

Arthur March died on April 17. When Erwin wrote to Hilde he began by quoting a verse from Rilke:

> Wir wissen nichts von diesem Hingehn, das
> nicht mit uns teilt. Wir haben keinen Grund,
> Bewunderung und Liebe oder Hass
> dem Tod zu zeigen, den ein Maskenmund
> tragischer Klage wunderlich entstellt.

We know nothing of this passing, which / sends us no message. We have no reason, / to show admiration or love or hatred / towards Death, which a face mask / of tragic sorrow strangely distorts.

Erwin wrote that one should try to fill the last days of a dying man with beauty rather than simply trying to prolong his life for a few days more in a hateful hospital bed. We ought to be allowed to carry him into a favorite spot under a canopy of cherry blossoms. Then one would give him a *Heuriger* (last year's wine) in two or three *Viertels*, as much as he wishes. Then we should begin to add more and more opium, until with a radiant look he happily goes to sleep in that passing of which we know nothing. Unfortunately we must conquer Religion before such a passing will be possible, for as Lucretius said, *Tantum religio potuit suadere malorum*. Erwin spoke with special feeling since, as he told Hilde, he did not think that his own half-health could last much longer, three or four years at the most. "But I am glad of the few years, for the world is very, very beautiful."

In May 1957, Schrödinger came close to dying. A case of grippe had developed into pneumonia in the right lung. Penicillin and streptomycin had no effect, and the left lung was also attacked. Dr. Schneeweiss maintained a vigil at his bedside in the Vienna apartment. A trial of terramycin and magnamycin was begun. By the evening of May 29, he was weakening, and the doctors offered little hope. On the next day, however, Ruth's birthday, the fever began to fall, and Polzer now thought there was some hope. Hoff came and was definitely optimistic – the patient must be allowed to sleep, sleep, and more sleep. There was one nursing sister by day and one by night, and Erwin began to complain about the day sister – an encouraging sign.

During his fever he had talked a lot of physics and a lot of nonsense, but sometimes he made sense: "How good that our house could be brought together here, and all our books that we waited for." A friend from Graz had saved many of his books and it had been a joy to see them again after twenty years. Some of the books were those he had kept from his father's library: the collected works of Grillparzer in ten volumes, the poems of Kleist, the novels of Tolstoy. Among the many books that he had bought himself were volumes by Schweitzer, Darwin, Freud, Rilke, Hofmannsthal, Dante, Schopenhauer, Nietzsche, Russell, and a great variety of texts and commentaries on eastern religions.

On May 31, the patient was out of immediate danger. This was the day that the order *Pour le Mérite* was officially conferred in Bonn on Schrödinger and on Lise Meitner. Lise was the second woman to receive the honor, the first had been Kathe Kollwitz. Erwin had been greatly pleased when he heard from Otto Hahn the news of this highest non-military decoration of the German government, restricted to thirty living persons. The fact that the two distinguished Jewish women were willing to accept any award from the Germans would have seemed strange to Einstein, who had said that any official contact with Germany made him feel unclean. He had refused all invitations to participate in German organizations, for example, writing to the Max Planck Gesellschaft: "The crime of the Germans is truly the most abominable ever to be recorded . . . The conduct of the German intellectuals – seen as a group – was no better than that of the mob." Schrödinger did not share these feelings, but as he so often said, he was an unpolitical person and he had not been personally affected by the Holocaust.

LAST PROFESSORIAL YEAR

Schrödinger's final academic year at the university was from 1957 to 1958, but he was relieved of formal teaching duties. His last scientific lecture was given on March 26 in the Main Hall of Physics Institute II to a joint meeting of the Austrian Physical Society and the Chemical-Physics Society. His subject was "Is perhaps also the Energy Principle only Meaningful Statistically?" The "also" referred to entropy which was known to have a statistical basis. In this lecture he returned to the subject of his inaugural lecture in Zurich of thirty-six years ago, and now brought forth some new arguments: (1) In classical mechanics, the specification of energy depends upon an integration constant, which is only an approximation that breaks down in the limit of very small systems. (2) The so-called "stationary" energy levels, the only ones admitted in quantum mechanics, can be so densely packed that they cannot be safely distinguished on account of the uncertainty principle. (3) An eigenfunction with sharply defined energy is not an adequate description of the state of an isolated system, because there are many such states which are non-equilibrium states.

Schrödinger calls an energy eigenfunction "dead," since *nothing can happen* in an isolated system whose wave function is an energy eigenfunction. In the Zurich lecture, he had been concerned with energy exchanges in the interactions between atoms and radiation, but now he was interested in developing an analogy between energy and entropy for large-scale systems.

Bruno Bertotti attended this meeting and stayed in Vienna for several weeks. Erwin was delighted at this opportunity for discussions with the young scientist whom he hoped might extend some of his ideas. Anny gave a midday dinner party for Bertotti and Victor Weisskopf (who had come from Geneva). She served soup with dumplings (*Griesnockerlsuppe*), roast loin of veal (*Kalbsnierenbraten*) with rice and vegetables, *Sachertorte* with whipped cream, and a special Mocha coffee. The three generations of physicists, aged seventy-one, forty-nine, and twenty-seven, are said to have had a lively discussion of the energy problem after this repast.

PROFESSOR EMERITUS

During the summer and fall of 1959, Erwin and Anny spent four happy months in the Tirol. They stayed in Alpbach until almost all summer visitors had departed. Lise Meitner was there for part of their stay. As the weather grew colder, however, his old bronchitis recurred and he sometimes had trouble getting to sleep at night. He would then get up and write long letters to friends, such as one to Synge which included a restatement of his feelings about wave mechanics:

With very few exceptions (such as Einstein and Laue) all the rest of the theoretical physicists were unadulterated asses and I was the only sane person left . . . The one great dilemma that ails us . . . day and night is the wave-particle dilemma. In the last decade I have written quite a lot about it and have almost tired of doing so: just in my case the effect is null . . . because most of my friendly (truly friendly) nearer colleagues (. . . theoretical physicists) . . . have formed the opinion that I am – naturally enough – in love with "my" great success in life (*viz.*, wave mechanics) reaped at the time I still had all my wits at my command (1926 at the age of 39) and therefore, so they say, I insist upon the view that "all is waves." Old-age dotage closes my eyes towards the marvelous discovery of "complementarity" (Niels Bohr). So unable is the good average theoretical physicist to believe that any sound person could refuse to accept the Kopenhagen oracle.

He calls complementarity "a thoughtless slogan" and adds "If I were not thoroughly convinced that the man [Bohr] is honest and really believes in the relevance of his – I do not say theory but – catchword, I should call it intellectually wicked."

Before returning to Vienna, Erwin and Anny traveled for three weeks in Italy, visiting Mantova, Cremona, Pacenza, Parma, Verona, and Venice. "All very beautiful if one did not become tired." One wonders why they did not spend the next few months in some sunny and warm southern place, but probably the need for excellent medical care made them return to the refuge of their home in wintry Vienna.

Schrödinger began to work intensively on a second section of his book *Meine Weltansicht*. He maintained an extensive correspondence, and his letters were always full of good spirits and new ideas. Occasionally he would comment on his ill health, but always in an objective and non-complaining way. He bore the burdens of the flesh with optimistic resignation.

He found great consolation in the old Indian "This Thou Art." As he explained to Bertotti: "It is of course not a physical but rather a metaphysical statement. It is so simple that it is impossible to explain it. It cannot be grasped by the intellect, but it may spring up in you on some occasion like a spark, and then it is there and will never really leave you, even though it is not a practical maxim to use every hour of your life." The poet of the Bhagavadgita sees himself "not only in his dear friend, wife or son, but also in the lion that attacks him furiously, the snake that is about to give his child the deadly blow . . . finally even in the KZ [concentration camp] officer who inflicts devilish torture on his prisoners." Thus at first there is *nothing moral* at all in the *Tat twam asi*, it is something between intellectual and intuitional. Only, if it were shared by the mighty (at least), then the world would be a paradise. However, to the Brahmin this is only a very welcome by-effect. The main point is, as he puts it, to have recognized the truth. Also, it will comfort him in the hour of death, just as the "viaticum" comforts the Roman Catholic. Einstein, when he once (a long time ago, in his fifties) was dangerously ill, was asked by a near friend: Are you not afraid of death? No, he said. I feel so intimately connected with the whole of the world, that I cannot all of a sudden drop out of it.

Schrödinger maintained an interest in many things besides science and philosophy, including even the Vienna physics institute. He wrote to Born early in 1960: "The ship of physics has been leaderless (here) ever since Hasenöhrl was killed. For thirty years the School was directed by my friend Thirring. Next semester his son takes over and all hell breaks loose. Hans was not gifted for theoretical physics. Walther is an excellent specialist."

In the spring of 1960, Erwin's respiratory symptoms seemed difficult to interpret and extensive tests were carried out. To the surprise of his doctors, evidence was found for recurrence of the pulmonary tuberculosis of forty years ago. He was ordered to begin a *Liegekur* with a course of tuberculostatic drugs. Thus in the late spring he found himself lying on a *chaise longue* on a sunny verandah in Alpbach with a view of snowcapped mountains, a virtual reconstruction of his situation in Arosa in 1921.

WHAT IS REAL?

During his *Liegekur* in Alpbach, Schrödinger completed the second part of *My World View*, which he called *What is Real?* In sixty pages he gave a final statement of his philosophy. He knew that his life was drawing to an end, and the book is written in an informal style, serious but undogmatic, with a number of homely examples to illustrate difficult points. It was not meant to be a philosophical treatise with logical and weighty arguments, but simply a translucent statement of his personal beliefs. Indeed, as he admits, it is not possible to give any rigorous demonstration of what he calls these mystical–metaphysical propositions.

23. Discussion with Lise Meitner at Alpbach

His first chapter is "The Reasons for the Abandonment of the Dualism of Thinking and Being or of Mind and Matter." The most cogent reason, he says, is that we are not able to explain how a material world can act upon an immaterial mind, or *vice versa*. To eliminate this intractable problem, we accept monism and reject dualism. Such a choice is also more economical of hypotheses and is thus encouraged by the threat of Occam's razor.

Having opted for monism, we must now choose between matter and mind. One of the earliest exponents of a purely material world was Democritus of Adbera, who taught that the soul is made of atoms of an especially fine, smooth, spherical, readily mobile kind. A noted advocate of the primacy of mind was Spinoza, who said that the all was one substance, with two attributes, extension and thought, and this unity he called God. Schrödinger admits that a *representation* of a real external world is essential to everyday life, but we do not need the existence of an object that is represented, indeed, we do not know what "to exist" would mean for such an object. He says that if we are to have only one realm, it must be the psychic since the psychic certainly exists (*cogitat–est*). An argument *against* this view has been stated by William James: consciousness or thought is not an *entity*, it is a *function* of a structure such as a human brain.

The theme of Chapter 2 is that we become aware of the commonality of the world only through language, which includes both verbal and non-verbal communication. This shared experience is a basic fact. The hypothesis of a common external world does not "explain" this fact – it merely restates it in different words. "Reality, existence, and the like, are empty words." Even if the hypothesis of a common external world is plausible, it still does not explain how we recognize its commonality. The next chapter explores "The Imperfection of Understanding": no science is ever exact or is ever completed.

After these preliminaries Schrödinger restates the "Identity Doctrine," the teaching of Vedanta that the underlying reality is a unity of Mind. In this last statement of his beliefs, he gives up any reliance on logical argument. "It is now necessary to admit that the considerations of this chapter are *logically* not meant quite so seriously as all that was previously discussed, but *ethically* much more seriously . . . Now I shall not keep free of metaphysics, nor even of mysticism; they play a role in all that follows." In brief, the meaning of Vedanta is that "we living beings all belong to one another, that we are all actually members or aspects of a single being, which we may in western terminology call God, while in the Upanishads it is called Brahman."

Only the inner vision and realization of the unity in Brahman will free the soul from the recurrent *Samsara* of reincarnation, worldly life, and death. "One must make the mystical doctrine one's own, understand it with the entire soul, and not only with lip-service."

Schrödinger calls this a "salvation through knowledge," and he seems to believe that it requires only intelligence and leisure for contemplation, apparently forgetting the role of ascetic practices. His studies of Vedanta must have made him aware that it prescribes a way of life and not merely a system of beliefs, yet in all his writings, including this last testament, he remains curiously aloof and impersonal. It is not known that his belief in Vedanta ever influenced his actions as distinct from his philosophical writings, and he also kept this belief scrupulously separated from his work in theoretical physics, including his interpretation of wave mechanics.

During his *Liegekur* Schrödinger also wrote a short (forty-page) autobiography *Mein Leben*. It gave most emphasis to his childhood and student days. The time of famine in Vienna after the First World War had left indelible traces in his memory and, although he would not have liked the word, in his conscience. Even now, in the last year of his life, he blamed himself for not doing more for his mother and father when they had lost their money. Actually he was financially helpless at that time, and from this impotence came his lifelong dislike for bankers and profiteers.

He wrote particularly to Bertotti, Born, Synge, and the poet and novelist Franz Csokor. His letters ranged over many subjects, from physics to literature and even politics. To Born he wrote a final denunciation of the probability interpretation of the wave-function:

Maxel, you know I love you and nothing can change that. But I do need to give you for once a thorough head washing. So stand still. The impudence with which you assert time and again that the Copenhagen interpretation is practically universally accepted, assert it without reservations, even before an audience of the laity – who are completely at your mercy – it's at the limit of the reputable . . . Have you no anxiety about the verdict of history? Are you so convinced that the human race will succumb before long to your own folly?

In his last letter to Born, on October 24, he mentioned that he had twice nominated for a Nobel prize the Viennese physicist Marietta Blau, who had discovered the "stars" in photographic plates caused by massive high energy particles in cosmic-ray showers. Blau was one of the few Jewish refugees who returned to Austria after the war. She lived in poverty since, in accord with the usual Austrian policy, she was offered no reparations. Walter Thirring offered help and Schrödinger managed to obtain for her the Schrödinger prize, which had been established by the Academy of Sciences, but this provided only a temporary support.

LAST DAYS

On October 20, Anny narrowly escaped death in the worst asthma attack she had ever experienced. By good luck there was a Red Cross doctor nearby, who brought oxygen that saved her from asphyxiation, and she was rushed to the hospital. She was destined to survive Erwin for over four years.

On November 9, Erwin decided to return to Vienna alone. There was no maid at the apartment and Erwin had considerable difficulty managing for himself. He was helped by Anny's sister Irmgard and by the couple who were caretakers of the building. He drew up a schedule for himself "For the Time When Anny is not Here," which ended with the words, "If I become really sick, naturally I must enter the hospital, that is clear." Anny worried about their separation as "joy and sorrow has bound us so closely together in the past forty-one years that we don't want to be separated during the few remaining years of our lives." They wrote loving letters to each other every few days.

Erwin became too weak to take care of himself and on December 2, he was brought to the Lainz Hospital where Polzer intended to make an examination of his lungs. He had a fine corner room with a view of the Lainz Zoo, but he protested so loudly and so furiously that he was transferred to the university psychiatric–neurological clinic. When he calmed down, he was much amused by this, exclaiming, "Now at last I am with Hoff." (Anny had spent many weeks under the care of the famous psychiatrist.)

By early December Anny was well enough to return to Vienna and she was able to spend several hours a day with Erwin. He asked to be brought home, saying, "I was born at home and I'll die at home even if it shortens my life. At the age of 73, I am not going to put up with forced nursing." In the week before Christmas, his condition became critical, with a

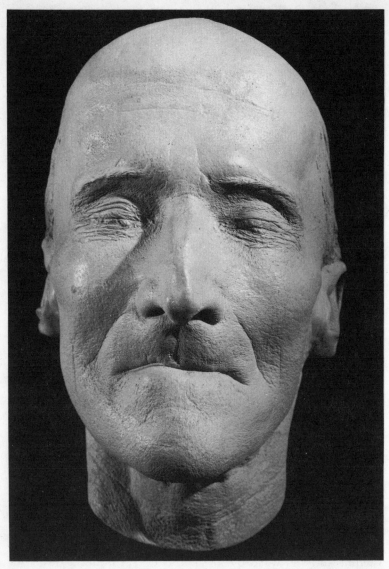

24. Death mask

urological problem complicating the heart disease. Anny held his hand for hours each day, leaving the bedside only for meals and sleep. He often said "Oh since I have you again, everything is good again."

The doctors finally agreed that he should be brought home in accord with his wishes, and he was put into his bed there at 10:00 a.m. on January 3. He was still conscious but he could no longer breathe without an oxygen apparatus. At 4:00 p.m. in the morning he became unconscious, but a doctor who immediately came gave him an injection which brought him back to consciousness. Anny sat by his bedside. He could no longer take nourishment. In the afternoon she gave him a sip of orange juice, and then heard his last words: *"Annichen, Du bleibst bei mir – auf das ich nicht hinunter stürze."* (Annikin, stay with me – so that I don't fall down.) Then he became still and seemed to sleep. Around 5:00 p.m. the doctor came and said that he could last only a few hours. At about 6:55 p.m. his pulse stopped – he was dead. Anny kissed him, stroked his hair, and took her departure from him.

Hans Thirring came immediately and took over the organization of everything. The coroner arrived and certified the official death and the body was taken to the Medicolegal Institute where two death masks were taken and also a casting of his hand. Dr. Holczabek, dean of the medical faculty, performed an autopsy. The cause of death was reported to be general ageing of the heart and arteries, with no specific pathology.

The body was transferred to Alpbach where it arrived exactly as the bell of the little church was striking noon on Saturday, June 9, 1961. There was some problem about burial in the churchyard since Erwin was not a Catholic, but the priest relented when informed that he was a member in good standing of the Papal Academy, and a plot was made available at the edge of the *Friedhof*. The ceremony at the graveside was simple but impressive. Hans Thirring gave a talk and Father Messner said the Lord's Prayer. Almost all the villagers of Alpbach were gathered around. The next day the gravemound was completely covered by snow.

Index